I0001058

Douglas Montague Gane

The Building of the Intellec

Douglas Montague Gane

The Building of the Intellec

ISBN/EAN: 9783337366162

Printed in Europe, USA, Canada, Australia, Japan

Cover: Foto ©berggeist007 / pixelio.de

More available books at **www.hansebooks.com**

THE BUILDING OF THE INTELLECT:

A Contribution towards Scientific Method in Education.

BY

DOUGLAS M. GANE,

AUTHOR OF ' NEW SOUTH WALES AND VICTORIA IN 1885.'

' Mens sana in corpore sano.'
JUVENAL.

LONDON:
ELLIOT STOCK, 62, PATERNOSTER ROW.
1897.

Dedicated.

WITH KIND PERMISSION,

TO THE

Rt. Honourable Sir JOHN LUBBOCK, Bart., m.p.,

F.R.S., D.C.L., LL.D.,

WHOSE VARIED ATTAINMENTS,

IN THOUGHT, IN SPEECH, AND IN ACT,

PRE-EMINENTLY EXEMPLIFY

THE AIM AND OBJECT

OF

ALL TRUE EDUCATIONAL METHOD.

PREFACE.

In the following pages I have attempted a brief abstract of some of the leading thought on Education.

I have sought not so much to give expression to any views of my own, as to present a systematic exposition of such opinions as are most deserving of attention in the works of others.

Not to diminish the force or authority of such opinions, I have thought it desirable to quote the arguments and conclusions of my authors literally, rather than convey them in language of my own.

Education being now regarded as a question of such vital importance, and opinions differing so widely as to its method, character and scope, a summary of the views of those best qualified as guides and teachers cannot fail to be of service to us in our efforts to arrive at something like unanimity of opinion.

In view of the promise of modern psychological inquiry to give Education a scientific basis, I have treated mental phenomena as the only sure guide to sound educational method. Every admission of new data, and verification of old, being an acquisition to the science they concern, I have ventured to submit for the approval of my readers several conclusions · at which I have arrived, on a consideration of the parallel presented by the mental evolution of the individual and the mental evolution of the species.

Though I have been careful not to disregard the advantages of Education as directed to the attainment of scholarship, I have preferred to treat it primarily as an instrument in the formation of character.

I have avoided detail, not only with the object of confining my remarks within moderate limits, but also because I consider that, for the general reader—for whom the present work is intended—such a subject should be treated only in its general bearings.

With such limitations, I have endeavoured to present a summary of views on the subject of Education that may prove serviceable to those entrusted with the care of the young, and who are brought face to face with the difficulties attending it.

Furthermore, Education — apart from its practical aspects, and its value as a means to the due appreciation of all that is great and noble in Literature, Art, and Science—is in itself attractive to every thinking mind. To the reader, therefore, who peruses the following pages with no more serious motive than the interest afforded by the subject, the present work may prove not undeserving of attention.

RICHMOND HILL,
July 21, 1897.

CONTENTS.

CHAPTER I.

SOME DEFINITIONS.

CHAPTER II.

A CONSIDERATION OF METHOD.

CHAPTER III.

THE DEVELOPMENT OF THE POWERS.

PAGE

CHAPTER IV.

THE CLASSICS.

CHAPTER V.

GRAMMAR—MODERN LANGUAGES—HISTORY.

CHAPTER VI.

MATHEMATICS—SCIENCE.

CHAPTER VII.

ART.

CHAPTER VIII.

THE BASIS OF MORALITY.

CHAPTER IX.

TEACHERS.

'Still, it must be allowed that learning does take away something, as the file takes away something from rough metal, the whetstone from blunt instruments, and age from wine ; but it takes away what is faulty ; and that which learning has polished is less only because it is better.'—QUINTILIAN.

INTRODUCTION.

THE picture of man given us by Sir John Herschell, when judged only with regard to his physical constitution, is in strange contrast with the nobility and power he has displayed when raised to the meridian of his intellectual greatness. ' No other animal,' says this distinguished author, ' passes so large a portion of its existence in a state of absolute helplessness, or falls in old age into such protracted and lamentable imbecility. To no other warm-blooded animal has Nature denied that indispensable covering without which the vicissitudes of a temperate and the rigours of a cold climate are equally insupportable; and to scarcely any has she been so sparing in external weapons, whether for attack or defence. Destitute alike of speed to avoid and of arms to repel the aggressions of his voracious foes, tenderly susceptible of atmospheric influences, and unfitted for the coarse aliments which the earth affords spontaneously during at least two-thirds of the year, even in temperate climates, man, if abandoned

I

to mere instinct, would be of all creatures the most destitute and miserable.'

In spite of these physical disadvantages, however, man, Sir John Herschell goes on to remark, 'is the undisputed lord of the creation.'

We will not now attempt to trace the order of development by which the great minds of the earth have been evolved from a source apparently so unpropitious. Suffice it to say that countless generations separate these extremes. The accumulated experience of mankind has grown in volume until, in the fulness of time, a Plato, a Leibnitz, or a Descartes has been born.

But if we are to seek an instance in illustration of the height to which the mind of man has been shown capable of development, we can perhaps find none more forcible than Aristotle. His genius was universal, and his mental achievements were destined for nearly two thousand years to form the beacon light of intellectual effort. 'His aim,' says Sir Alexander Grant, 'was to produce what we should call an encyclopædia of all the sciences. . . . He began from the very beginning, and framed his own philosophical or scientific nomenclature; he traced out the laws on which human reasoning proceeds, and was the first to reduce these to science, and to produce a logic. He wrote anew "Metaphysics," "Ethics," "Politics," "Rhetoric," and "The Art of Poetry"; and while these were still on the stocks, he was engaged in founding on the largest scale the physical and natural sciences, especially natural philosophy, physiology under various aspects (such as histology and anatomy, embryology, psychology, the philosophy of the senses, etc.), and, above all, natural history.'*

And the result of this vast learning and industry the same author sums up in the following words : 'In the history of European thought and knowledge, down to the period of the revival of letters, the name of Aristotle was without a rival, supreme. . . . His works may almost have the credit of having saved men from relapsing into barbarism.'†

But we are not confined to ancient Greece for instances of vast mental capacity. All times and all places within the limits of civilization have given birth to prodigious intellects. For example, Leonardo da Vinci, Sidney Colvin says, 'excelled in almost every honourable human attainment, the commercial and political excepted. . . . In the fine arts he was the most accomplished painter of his generation, and one of the most accomplished of the world, a distinguished sculptor, architect and musician, and a luminous and pregnant critic. In inventions and experimental philosophy he was a great mathematician and engineer, an anatomist, a botanist, a physio-

---

* 'Encyclopædia Britannica.'
† Sir A. Grant's 'Aristotle.'

logist, an astronomer, a chemist, a geologist and geographer—an insatiable and successful explorer, in a word, along the whole range of the physical and mathematical sciences when most of these sciences were new.'*

To turn to our own countrymen, in the poet Gray we find a memorable instance of comprehensive knowledge. Of his attainments Matthew Arnold writes on the authority of the poet's friend, Temple: ' Mr. Gray was, perhaps, the most learned man in Europe. He knew every branch of history, both natural and civil; had read all the original historians of England, and France, Italy, and was a great antiquarian. Criticism, metaphysics, morals, politics, made a principal part of his study. Voyages and travels of all sorts were his favourite amusements; and he had a fine taste in painting, prints, architecture, and gardening.'

Yet vast as were the attainments of these men, we are justified in saying that their achievements by no means represent the limit of intellectual greatness. Moreover, it is consoling to reflect that, boundless as is the honour we pay to the intelligence of predecessors such as these, we are not by reason of that ' condemned to stupidity' in our own age. ' It is sufficient for encouragement in study,' says Quintilian, ' to know that it is not a law of Nature that what has not been done cannot be

* ' Encyclopædia Britannica.'

done.; and in the second, that everything great
and admirable had some peculiar time at which
it was brought to its highest excellence. . . .
Whatever is best had at one time no existence. . . . The advancement of the arts to
the highest possible excellence would be but an
unhappy service to mankind, if what was best
at any particular moment was to be the last.'

At the same time, it is not to be denied that
the appearance of intellectual giants such as we
have named is due to the co-operation of many
favouring circumstances seldom found in combination. Locke has given it as his opinion
that, of all the men we meet with, nine-tenths
of them are what they are, good or evil, useful
or otherwise, according to their education.

Without doubt education can do much,
but it cannot altogether transform what is
born in us; for points of nature, Lord Bacon
observes, are without our command. 'If it
deserve to be considered,' he remarks, 'that
there are minds which are proportioned to
great matters, and others to small, doth it not
deserve as well to be considered that there are
minds proportioned to attend to many matters,
and others to few? So that some can divide
themselves, others can perchance do exactly
well, but it must be but in few things at once:
and so there cometh to be a narrowness of
mind, as well as a pusillanimity. And, again,
that some minds are proportioned to that which

may be despatched at once, or within a short
return of time; others to that which begins afar
off, and is to be won with length of pursuit.'

In regard to points of fortune, which Lord
Bacon regards as an obstacle to education
equally beyond our powers of control, Ignatius
Loyola declared, and not without some near
knowledge of the matter, 'that in absolute
penury the pursuit of science cannot easily
subsist, and the culture of the mind is impeded
by the duties of providing for the body.'

But the whole value of education does not
lie in the possibility of arousing prodigious
intellects to activity. Education is necessary
to the average man, both for forming his
character and for fitting him to fulfil creditably
his mission in life.

He must be taught to think, and so to
reason; for 'the real freedom of a rational being
consists in an ability to regulate his conduct by
reason.' But to reason accurately he must
have knowledge, for to reason from opinions,
and not from facts, is to lay himself open to
false conclusions. Hence Heracleitus says:
'The mass of men live as if each had a
private wisdom of his own.'

Prejudice is the natural consequence of
reasoning from opinions rather than from facts;
hence our prejudice is generally proportionate
to our ignorance. From this it will be seen
that, in our dealings with our fellow-men, the

more knowledge we possess, the fairer are our judgments likely to be. Again, if our knowledge be enlarged, and our reasoning powers matured, may we not hope to escape some of those calamities that, according to Seneca, attend the guidance of the uncultivated judgment ?

'So great are our mistakes in the true estimate of things,' this philosopher observes, 'that we have hardly done anything that we have not had reason to wish undone, and we have found the things we feared to be more desirable than those we coveted ; our very prayers have been more pernicious than the curses of our enemies, and we must pray again to have our former prayers forgiven. Where is the wise man that wishes to himself the wishes of his mother, nurse, or his tutor, the worst of enemies with the intention of the best of friends ?'

In our daily life we have not only to form conclusions, but are at times called upon to give our reasons for them. 'Men's minds,' says Dr. Whewell, 'are full of convictions which they cannot justify by connected reasoning, however reasonable they may be. Nothing is more common than to hear persons urge very foolish arguments in support of very just opinions, and what has been said of women is often no less true of the sex which pretends to have the more logical kind of head : namely,

that if they give their judgment only, they are
not unlikely to be right, but if they add their
reasons for it, those will most probably be
wrong.'

Hence one inducement of education to a man
is 'to enable him to render a reason for the
belief that is in him.'

But knowledge is necessary for the main-
tenance of our health and the preservation of
our fortune. For though, remarks Herbert
Spencer, 'Nature has provided efficient safe-
guards to health, lack of knowledge makes
them in a great measure useless.'

We cannot of course hope, nor should we
attempt, to undertake those offices for which
years of study and practice alone can qualify
us, and which even then we can fill with
only partial success. But we can acquire
such a knowledge of their general theory
as will save us from imposition, and will give
us greater facilities for distinguishing merit
and detecting incompetency. Moreover, we
shall be contributing to the value of the
advice we seek if we approach our adviser
with a cultivated mind. For, says Huxley,
in speaking of the art of medicine, 'it is, I
think, eminently desirable that the hygienist
and the physician should find something in the
public mind to which they can appeal; some
little stock of universally-acknowledged truths,
which may serve as a foundation for their

warnings, and predispose towards an intelligent obedience to their recommendations.'

These remarks apply with equal force to the study of law. In fact, Mr. Justice Stephen, in his ' Commentaries on the Laws of England,' thus recommends to his countrymen a better knowledge of the laws to which they have to render obedience. He remarks : ' The subject (the English law) is one to which no class of readers in the realm can be indifferent ; for it is incumbent upon every man to be acquainted with those laws at least with which he is immediately concerned, lest he incur the censure, as well as the inconvenience, of living in society without knowing the obligations which it lays him under. But it ought to have peculiar attractions for men of liberal education and respectable rank. These advantages are given them, not for the benefit of themselves only, but also of the public ; and yet they cannot in any scene of life discharge properly their duty, either to the public or to themselves, without some degree of knowledge in the laws.'

But education offers a further inducement. If duty or utility fail to attract, there yet remains pleasure. ' Shall we,' says Ernest Chesneau, ' appeal to a final inducement, less lofty, but not less powerful ? May we not invoke the spirit of dilettantism—the gratification of refined personal tastes ? Is it not a fact,' he continues, ' that the more open and sensitive the mind is to intellectual impressions, and the more instinct with moral vitality, the greater are its opportunities of unexpected enjoyment?'

Indeed, our sources of pleasure are greatly multiplied. Objects of interest arise in places which, were the mind less enlightened, would be thought destitute of attraction. And as education advances, the faculties grow keener, and the mind awakens to the consciousness of the vast fields of knowledge that present themselves for its exploration. A limited knowledge of astronomy discloses in the firmament above us innumerable wonders that enable us to see in a starlight night something more than a mere condition of fine weather. A moderate acquaintance with the principles of art will give a delight to our visit to a picture-gallery that we could not otherwise have experienced ; and an elementary knowledge of architecture will enable us to find in a cathedral a volume written in a language which only our ignorance precluded us from reading. In a country walk, too, how many things come into notice if we have but knowledge enough to call the flowers by their names, and how our interest in them will expand if we are but slightly intimate with their structure and classification !

But education does not merely afford pleasure : it does more. It affords a refined pleasure —a pleasure which, in the words of Huxley, ' is neither withered by age, nor staled by custom, nor embittered in the recollection by the pangs of self-reproach.'

And this is an important aspect of education; for, says the same author, ' a pleasure-loving character will have pleasure of some sort ; but if you give him the choice, he may prefer pleasures which do not degrade him to those which do.'

And the enjoyment reaped from education may yet reach a fuller height ; for ' the pleasure derived from knowledge is pure when it is withdrawn from action, that is, from toil, and enjoys the calm contemplation of itself.'*

' How great a thing it is,' says Rabanus, ' to approach in spirit to the heavens—to explore all their supernal mechanism by rational investigation, and by lofty, intellectual insight to observe anywhere and everywhere the veiled secrets of their vast greatness !'

Thus is the mind brought back to religion, which a little or superficial knowledge, observes Lord Bacon, may have inclined to atheism.

* Quintilian's ' Institutes.'

CHAPTER I.

SOME DEFINITIONS.

'Education must come before philosophy.'—PYTHA-GORAS.

'Only let it be accepted as a cardinal law of education that before it can do any profitable work the mind must have material to work upon.'—E. E. BOWEN.

'The saying that a little knowledge is a dangerous thing is, to my mind, a very dangerous adage. If knowledge is real and genuine, I do not believe that it is other than a very valuable possession, however infinitesimal its quantity may be. Indeed, if a little knowledge is dangerous, where is the man who has so much as to be out of danger?'—HUXLEY.

'And the error is, that after the distribution of particular arts and sciences men have abandoned universality, or philosophia prima, which cannot but cease and stop all progression.'—BACON.

IT would be difficult to define education in such a way as to include the multitude of ideas that have been formed concerning it, differing more or less as they have done with each epoch of the growth of civilization. Education in its larger sense, John Stuart Mill has said, is one of the most inexhaustible of all topics, and one

of the subjects which most essentially require
to be considered by various minds, and from
a variety of points of view. But impossible
though an accurate definition may be, there are
certain fundamental notions with regard to its
aim and object about which there is little or
no difference of opinion; and these are, that
it should comprise the development of the
faculties, the acquisition of knowledge, and the
enlargement of the power of using knowledge
profitably, which is skill. For 'the difference,'
observes Dr. Arnold, ' between a useful education and one which does not affect the future
life rests mainly on the greater or less activity
which it has communicated to the pupil's mind,
and whether he has learned to think or to act,
and to gain knowledge by itself, or whether he
has merely followed passively as long as there
was someone to draw him.'

In Greece, Dr. Donaldson tells us, the chief
aim of education was the realization of capacity,
and not the furthering of acquisition; for ' in
the best period of the history of that country,'
remarks Mr. Perry,* ' there was no such thing
as learning or learned men, but only thinkers
and actors, whose minds developed themselves
in their natural shape.'

The training of the intelligence of children
is, it is said, mainly concerned with the development of mental habits. ' All that a school can

* 'Greek and Roman Sculpture.'

teach,' observes Dr. Wiese,* 'beyond a certain
small stock of knowledge is the way to learn.
It is a lamentable misconception of that most
important maxim to suppose that a liberal
education can have any other end in view than
to impart and exercise power to be used in
after-life.' Though, in part, we agree with this
definition, we take exception to the limitations
it imports. Let us examine the question more
closely.

Education realizes its final aim in action.
We learn in order that we may do. We master
each science that we may exercise its corresponding art. 'A science,' says Sir James
Sawyer, 'has to do with knowing, an art with
doing. You must learn each science in order
to understand its correlative art; you must
practise each art if you would illustrate and
appreciate the principles of its science.' The
art of medicine is based upon the knowledge
of the human body. The art of thinking is
based upon the knowledge of the material of
thought; and the art of right thinking is based
upon the knowledge of what is right and estimable in the material of thought. There is, in fact,
an art of conduct, and it can be exercised only
after the knowledge is first acquired of what is
good and noble in conduct. We cannot do right
deliberately until we first *know* what is right.

If, then, our schools teach but a small stock

* 'German Letters on English Education.'

of knowledge, it may satisfactorily explain the not uncommon want of moral rectitude, and the coarse tastes that distinguish many of our 'expensively-educated youths.' The period for acquiring knowledge is deferred too long. The school years are devoted to learning how to learn, and, when they are passed, the instrument that it has taken so many dull hours to sharpen is laid aside, its power unrevealed. Is it not lamentable that our school work, expressly dedicated as it is to give the means of acquiring knowledge, should in the final result prove so unproductive?'

· But what is this power that Dr. Wiese tells us our school training should give us for use in after-life? Is it not the power to be derived from the acquirement of certain mental habits —habits of seeing and knowing, habits of attention, memory, and judgment? Admittedly, these are but a means to an end. For why do we observe, except it be, through the medium of the senses, to bring new facts to the mind? What virtue is there in concentration, if it be not to render our knowledge of those facts more thorough and precise? And do we not treasure the faculty of memory for its power to preserve the facts for our future use? and in the operations of the judgment do we not put to use our store of accumulated facts for all the varied purposes of life?

It is manifest, therefore, that we develop the

faculties that we may acquire knowledge ; and we acquire knowledge that we may perform the numerous duties of life with wisdom, temperance and precision.

Of the duties of life, some are general, others special. Some are imposed upon the individual in his capacity of a man, a rational being ; others affect the individual in his special capacity, say as a statesman, a doctor, or a lawyer. The knowledge that is necessary for the fulfilment of general duties should form the subject of school and the very earliest teaching. The knowledge that is required for the performance of special duties should be acquired when school-days are over. We will return to this later.

The objects of education accordingly arrange themselves under two heads : namely, the development of the mental faculties, and the acquisition of such knowledge as is necessary for the building up of the rational man.

Habits of mind are acquired by virtue of things done, not things known. Observation becomes a habit by the practice of observation; attention by the practice of attention; reasoning by the practice of reasoning. Knowledge may guide the working of the faculties, but for developing them, it is no substitute for their active use. For instance, knowledge acquired direct from things stimulates observation; the same knowledge acquired from books tends to check it. The habit of thought comes from the practice of thinking ; but heads ' replete with thoughts of other men ' do not think, or think very little.

The subjects of the educational course are, it is claimed, chosen firstly with a view to their power to exercise the mental faculties, and secondly from a regard to the value of the knowledge they impart. A moment's consideration will show that the former is the prevailing factor in the choice. The Classics, it is said, furnish a mental gymnastic of the highest order ; and they impart the most valuable knowledge when access to the literatures has been attained. But access to the literatures, in the cases of ninety-nine students out of every hundred, is not attained, a fact that is well recognised, and yet the Classics are retained in the school curriculum.

In later chapters we shall treat more in detail of the qualities of the main subjects of study. Suffice it now to say that a school subject that does not meet both the requirements we have named, that does not store as well as exercise the mind, lacks one of the vital qualities of a school study, and should not be retained in the course to the exclusion of one that does.

But unlike the Classics, as they are proved to be to the majority, there are studies which, while they develop the faculties, at the same time impart knowledge. The question then

arises, What is the value of the knowledge?
Between various disciplinary studies in which
the quality of the knowledge imparted differs,
that study by which the most valuable knowledge is given should be the subject of the first
choice.

Though, as we have shown, it is desirable
in school years to limit the training to the
development of mental power and the acquisition of general knowledge, yet various exceptions are made, by common consent, to this rule,
and the pupil receives particular instruction in
some special branch of knowledge, with a view
to pursuing the art to which it is the introduction. This arises in the case of painting and
music.

But though exceptions are made in these two
cases, it is not desirable to extend them. 'If
the young,' said Aristotle, 'had to learn everything that might be useful to them in after-life,
they would have to descend to cookery.' 'Un-
discerning parents,' says Herbart, 'may drill
their sons and daughters according to their
tastes ; they may lay all kinds of varnish on
the unpolished wood, which in years of inde-
pendence will be roughly rubbed off, but not
without pain and injury.'

Spencer maintains that accomplishments,
employing as they do the leisure part of life,
should occupy the leisure part of education.
But if that prevailed, many a brilliant painter

or musician might be lost to us for lack of serious rudimentary training in early life. In any case, it is obvious that ' we import into the legitimate province of the teacher that which does not belong to it when we crowd a mass of multifarious acquirements into the period of life allotted to the growth and improvement of our reasoning powers and our physical energies.'*

As we have shown, the period of life allotted to school and college training is not that in which the student should receive instruction in the particular business to which he will apply himself in after-life. The two courses of train-.ing should be kept distinct, and the one superposed on the other. If the business training be allowed to usurp the time that should properly be allotted to the general training, both will suffer : the general training by reason of its lost opportunity, and the business training by reason of the inadequate basis on which it is built.

'Men are men,' urges John Stuart Mill, 'before they are lawyers or physicians, or merchants or manufacturers; and if you make them capable and sensible men, they will make themselves capable and sensible lawyers or physicians. What professional men should carry away with them from a university is not professional knowledge, but that which

* Dr. Donaldson.

should direct the use of their professional knowledge, and bring the light of general culture to illuminate the technicalities of a special pursuit.'

To the correctness of this view Dr. Donaldson bears additional testimony when he quotes one of the State Reports on Education, bearing the signatures of Mr. T. B. Macaulay, Lord Ashburton, the Rev. H. Melvill, Professor Jowett, and Mr. Shaw Lefevre. The report runs: 'We believe that men who have been engaged up to twenty-one or twenty-two in studies which have no immediate connection with the business of any profession, and of which the effect is merely to open, to invigorate, and to enrich the mind, will generally be found in the business of every profession superior to men who have, at eighteen or nineteen, devoted themselves to the special studies of their calling.'

The question is a vital one, and its importance should be more widely recognised. To illustrate our meaning more fully, let us briefly apply our general proposition to the cases of particular pursuits. Three examples will perhaps suffice :

In regard to the study of medicine, Sir James Sawyer, in one of his hospital addresses remarks: 'At a time when some of you, perhaps, may be inclined to over-estimate the relative importance of a purely professional as compared with a general training, pray let me tell you that you are not fitly prepared to come here unless you have been well grounded in the rudiments of a liberal education. . . . Depend upon it, the progress you may make in your studies here, no less than the position you will occupy hereafter in public and professional estimation, depends very greatly on the degree and extent of mental cultivation to which you have already attained.'

In reference to the pursuit of art, Ernest Chesneau writes : ' A great artist is an impossibility without a general education, and a man who has no artistic culture, however superior he may be in other respects, lacks an instrument which is indispensable to his complete use of life.'

As to the training needful for the artisan, Huxley says : ' In my judgment the preparatory education of the handicraftsman ought to have nothing of what is ordinarily understood by "technical" about it. The workshop is the only real school for a handicraft. The education which precedes that of the workshop should be entirely devoted to the strengthening of the body, the elevation of the moral faculties, and the cultivation of the intelligence, and especially to the imbuing the mind with a broad and clear view of the laws of that natural world with the components of which the handicraftsman will have to deal.

And the earlier the period of life at which the handicraftsman has to enter into actual practice of his craft, the more important is it that he should devote the precious hours of preliminary education to things of the mind which have no direct or immediate bearing on his branch of industry, though they be at the foundation of all realities.'

But when the period of academic life is past, and the scholar is thrust on the world with his mind rich in the capacity to accumulate knowledge, he may be tempted to pursue his 'thought-building' into all departments of learning. But the exigencies of life are such, that, in order to fit himself for the career he determines to make his own, he will have to be content, for a time at least, 'in order to know a little well, to be ignorant of a great deal.'

'You cannot learn everything,' says Froude: 'the objects of knowledge have multiplied beyond the power of the strongest mind to keep pace with them all. You must choose among them, and the only reasonable guide to choice in such matters is utility.'

But though we cannot hope to excel in numerous studies, the attainment of due proficiency in any one study does not preclude the possibility of thorough general culture. On the contrary, a high standard of general culture will contribute to proficiency in the special pursuit. ' Let this be a rule,' says Lord Bacon, ' that all partitions of knowledges be accepted rather for lines and veins than for sections and separations, and that the continuance and entireness of knowledge be preserved. For the contrary thereof hath made particular sciences to become barren, shallow and erroneous, while they have not been nourished and maintained from the common fountain.'

Let it be, moreover, remembered that, for the correct estimate of life as a whole, and for the due adjustment of a man's life to those of his fellows—in other words, for the preservation of character—it is necessary that his mind should not be narrowed by the prejudices peculiar to his particular class or pursuit. It was a complaint of Heracleitus that men sought truth in their own little worlds, and not in the great and common world. If we are to have wide views, we must have comprehensive knowledge. It is this alone, ' the perception of similarity and agreement, by which we rise from the individual to the general, trace sameness in diversity, and master, instead of being mastered by, the multiplicity of nature,'* that gives us the deeper wisdom—that wisdom which, Lord Bacon says, ' endueth men's minds with the true sense of the frailty of their persons, the casualty of their fortunes, and the dignity of their soul and vocation.'

* Froebel.

CHAPTER II.

A CONSIDERATION OF METHOD.

'Order and constancy are said to make the great difference between one man and another. This I am sure: nothing so much clears a learner's way, helps him so much on it, and makes him go so easy and so far in any inquiry, as a good method.'—LOCKE.

Of Loyola's personal studies : 'As it was, they were as good as if they had never begun, or somewhat worse. He had gone about them the wrong way. Whatever solidity of learning he had kept objectively in view, something else, equally important with solidity, had been unwittingly omitted. That was a good method. Logic, philosophy, and theology, all taken up together, and with such compendious haste, now went together in his mind like a machine out of joint, and his speed was nil.'—REV. THOMAS HUGHES.

Of the studies of Leibnitz : 'So early, then, as his twentieth year we are led to recognise in Leibnitz's life and training peculiarities which are of importance to his philosophy. Few men began to build so early, few laid so broad a foundation ; no mistakes were made in construction, no doubts had to be overcome, no time was lost in aimless search. Every step he took was onward, every experience increased the means he already possessed, every stone which the builder took up was placed in the right position. No conclusions were arrived at hastily, no part of the edifice was finished prematurely. In harmony with itself, and without loss of evenness and balance, the work progressed steadily, a true example for

the historical student of that process which Leibnitz
himself afterwards conceived to be the explanation of the
world—viz., the individuality of everything real, and the
harmony of all things.'—JOHN THEODORE MERZ.

A STATEMENT of Aristotle's that has since
been generally allowed is that the development
of the body should precede the development
of the mind. ' A sound mind in a sound body ':
these combined qualities should be the constant
aim of those who have the care of children
entrusted to them. Health and vitality are the
first conditions of success in life, and the
student will have enriched his mind to small
purpose if he lack the physical strength to
fight his way in the battle of life.

But though the plant should be well grown
before the bud be induced to appear, yet in
the education of children the wholesome fear of
premature development of the mind must not
be allowed to unduly hinder the child's intellectual growth, for, observes Lord Bacon, ' the
culture and manurance of minds in youth hath
such a forcible (though unseen) operation as
hardly any length of time or contention of
labour can countervail it afterwards.'

In the Kindergarten system Froebel utilizes
the earliest years in imparting rudimentary
impressions to the child's mind, contending
that whatever is then gained is an acquisition
to youth. ' Whatever a boy has to learn,'
remarks Quintilian on this head, ' may he not

be too late in beginning to learn. Let us not, then, lose even the earliest period of life, and so much the less as the elements of learning depend on the memory alone, which not only exists in children, but is at that time of life most tenacious.' No age suffers less from fatigue, he continues, and by way of illustration points to the endurance shown by children in crawling on their hands and knees, and subsequently by their constant play and incessant activity; and he alludes to the ease with which children learn their mother-tongue, though no one incites them to learn—a great contrast, he observes, to the difficulty experienced in the same pursuit by the purchased slaves.

One of the greatest problems in connection with education is to determine what is best suited to the child's mind at particular periods of its growth. 'God does not cram in or ingraft,' says Froebel; 'He develops the smallest and most imperfect thing in continuously ascending stages and in accordance with eternal laws grounded in and developing from the thing's own self.'

We may safely affirm that the practice now so prevalent, arising from the overstrained system of competitive education, is not in harmony with these laws. 'It is well known,' says Matthew Arnold, 'that the cramming of little human victims for their ordeal of competition tends more and more to become an industry with a certain class of small school-masters ;' and he continues : ' The foundations are no gainers, and nervous exhaustion at fifteen is the price many a clever boy pays for over-stimulation at ten. . . . You can hardly put too great a pressure on a healthy youth to make him work between fifteen and twenty-five ; healthy or unhealthy, you can hardly put on him too light a pressure of this kind before twelve. . . . To put upon little boys of nine or ten the pressure of a competitive examination for an object of the greatest value to their parents is to offer a premium for the violation of Nature's elementary laws, and to sacrifice, as in the poor geese fatted for Strasburg pies, the due development of all the organs of life to the premature hypertrophy of one.'

But whatever the laws of mental growth may be, our efforts to comprehend them are met with greater difficulties than are perhaps at first apparent. We aspire to know what is funda-mental, what is general, and we meet with only what is special; and with our present knowledge we are unable to calculate the one from the other. The child lives his life among adults who, in their endeavours to mould him according to their particular ideas, are continually diverting the natural development of his personality. His instincts are checked, and the expression of his characteristic thought is modified to suit the taste and sensibility of the age. The simplicity and logic of his language is complicated by our penchant for irregularities, and his vanity and love of showing off are curbed in view of the canons of good behaviour. In matters of art, he adopts the taste and judgment of his elders ; and the more cultivated his surroundings, the more quickly will they modify his original native tendencies.

We draw attention to these few examples of outside influence on the mental growth of the child not to call in question their wisdom as such. We merely present them in proof of the fact that the phenomena of child-life as lived in our midst afford no available data from which to ascertain the fundamental laws of the development of the human mind. In fact, the child is more often the modified expression of his guardian's individuality than the faithful portrait of his own.

If, then, we can derive no satisfactory data from childhood itself, where are we to seek? Within the last few years interest has been aroused in the study of comparative embryology by the remarkable resemblances that are seen to exist between the embryos of the human and animal species in the early stages of their growth. · 'The eggs or germs,' says Mr. Clodd,* 'from which all organisms spring are, to outward seeming, exactly alike; and this likeness persists through the earlier

* ' The Story of Creation.'

stages of all the higher animals, even after the form is traceable in the embryos.' But not only in the embryos do we detect this curious correspondence, for it is found that the human fœtus reveals at successive stages of its growth features that are peculiar to, and in the order of their appearance mark the evolution of, the animal kingdom anterior to man. ' Thus,' remarks the same author, ' does the egg from which man springs, a structure only one hundred and twenty-fifth of an inch in size, compress into a few weeks the results of millions of years, and set before us the history of his development from fishlike and reptilian forms, and of his more important descent from a hairy, tailed quadruped.'

The positive results that have attended the inquiries of the embryologist are destined to bear yet more important fruit. The resemblance between the child and primitive man has been the subject of frequent remark,* but it is only of late that the matter has received serious consideration. The labours of those who have worked in this field of inquiry would, however, appear to justify the conclusion that the parallel that is established between the growth of the human fœtus and the evolution of the animal world admits of extension ; and

* Professor Sully, in his delightful book ' Studies of Childhood,' gives abundant illustration of this remarkable fact. Frequent allusion, too, is made to it in Professor Tylor's ' Primitive Culture.'

that, as the result, the child of civilized parents
and the primitive community in their journey
upward take the same course, the main difference being the swiftness of the one, the slow-
ness of the other. ' This same evolutional
point of view,' says Professor Sully, ' enables
the psychologist to connect the unfolding of
an infant's mind with something which has gone
before—with the mental history of the race.
According to this way of looking at infancy,
the successive phases of its mental life are a
brief résumé of the more important features in
the slow upward progress of the species.'

The parallel between the evolution of the
child and the evolution of the race is no doubt
in its general features satisfactorily established.
But it would be premature to claim from the
conclusions we are as yet justified in drawing
that we are therefore in possession of a final
solution of the difficult question of education.
Herbert Spencer, it is true, goes so far as to
say that 'the education of the child must accord
both in mode and arrangement with the education of mankind, considered historically.' But
his dictum we are inclined to think is one
of those daring hypotheses that Mr. Starcke
declares are the common fate of every dawning
science in its approach to truth. In the first
place, we have no definite knowledge of the
mode and arrangement of the education of mankind, considered historically. We have histories

of particular peoples setting forth more or less
fully the evolution of their national characteristics and aptitudes, marked by increasing
vagueness and uncertainty the farther back we
go in point of time, and leaving the long pre-
historic period in impenetrable mist. We have
the records of travellers revealing to us curious
facts regarding existing savage communities.
Many, no doubt, are accurate, but others require
corroboration before they can serve for scientific
use. Moreover, valuable as these observations
are when verified, they are usually disconnected,
' sawdustish,' to use a telling expression of
Carlyle's, and are utterly valueless for scientific
purposes until multiplied and arranged. With
histories that ignore the infancies of nations,
and records of existing primitive communities
that, taken in the mass, are in every way in-
adequate, it will be manifest that we are not
yet in a position to determine what is fundamental in mental evolution, and what is special
or accidental. True as it may be that the
life of the race represents the life of the individual microscopically enlarged, it is not equally
true that the life of any given nation represents
the life of any given individual. Nations differ
in every way as much as individuals. Greece,
for instance, as we know, began her memorable
career 'in the gorgeous clouds of fancy and
feeling.' Rome, on the other hand, was virtually
without a mythology of its own ; all that their

poets have to say with regard to the creation of the world and the origin of the gods is, without exception, borrowed from the Greeks.*

What we require to know, in investigating the laws of mental growth, is what is common to communities as a whole, what is special to each; what are the universal qualities of the human mind, and what are 'individual, national, or even racial, distinctions.'† To attain this we must carry our researches over a wide field of observation. Our prime sources of information are history and ethnology. But it is only by the method of comparative history and comparative ethnology that we can hope to extricate the fundamental character of the species from the acquired characters of nations.

When we have done this, we shall have gone far towards acquiring the true, natural system of education. We shall no doubt by this means learn that the periods of development have a well-assigned order, and that the difference in communities, as in individuals, does not lie in the order of procession of these periods, but in the intensity of them. In studying nations whose careers are esteemed memorable, we shall mark, where records exist, the characteristics of their infancy and youth, and thus detect the essential features that, with peoples as with men, are prophetic of future greatness.

* O. Seeman, 'The Mythology of Greece and Rome.'
† Tylor's 'Primitive Culture.'

We may further note the influences that helped or marred their progress. On the other hand, we shall mark, and so be forearmed against, those undesirable traits that by reference to this intelligible key will stand declared as evil tendencies.

Though, as we have stated, the mental evolution of any one race affords no sufficient clue to what is permanent and general in mental evolution, it may be useful and instructive to briefly sketch the mental history of some nation the records of whose descent are sufficiently precise and comprehensive. We naturally turn to the ancient Greeks. We make this selection not only on account of the comparative fulness of its records and the exhaustive criticism that has been applied to their investigation, but by reason of the phenomenal brilliance that marked the various periods of its intellectual growth. If the life of the race represents the life of the individual microscopically enlarged, our vision in viewing the latter will, through such a medium, be materially assisted. But how much more will this be the case if we choose for our inspection a race the features of which are in themselves of more than ordinary distinctness. Moreover, in taking as our instance the ancient Greeks, though we do not claim through them to give an exposition of the essential features of mental growth, we do hope by this choice to give the leading characteristics of mental growth in its most vigorous manifestation.

It is well known that the infancy of ancient Greece was marked by the extraordinary display of its imaginative faculty.* Most, if not all, nations, says Mr. Grote, have had myths, but no nation except the Greeks have imparted to them immortal charm and universal interest. Of all the authentic depositories of these early myths that have come down to us from the very era of their birth, but two remain—the 'Iliad' and the 'Odyssey'; but these two, says the historian above quoted, 'are quite sufficient to demonstrate in the primitive Greeks a mental organization unparalleled in any other people, and powers of invention and expression which prepared, as well as foreboded, the future eminence of the nation in all the various departments to which thought and language can be applied.'

The literature of this period was metrical, prose composition being a much later development. Writing was not general, the poems being committed to memory and handed down by oral transmission. They were sung or chanted at the national festivals, and were intended to appeal to the emotions of the hearers. The stories they embodied were regarded as true 'from their full conformity with the predispositions and deep-seated faith of an uncritical audience.' They appealed to a condition of morality that had its seat in the feelings, not in the judgment. Moreover, the poets were the sole instructors of youth. 'Down to the generation preceding Sokratês,' says Mr. Grote, 'the poets continued to be the grand leaders of the Greek mind. Until then nothing was taught to youth except to read, to remember, to recite musically and rhythmically, and to comprehend, poetical composition. The comments of preceptors addressed to their pupils may probably have become fuller and more

---

* The introduction of this historical summary into a work that professedly deals with education may seem to some of our readers a tiresome departure. We may say that the inferences we have drawn from it are intelligible without its previous perusal. On the other hand, it may be useful to those who wish to test for themselves the value of our conclusions.

instructive, but the text still continued to be epic or lyric poetry.'

The first trace we have of the emancipation of judgment from feeling is in 'the naked, dogmatical laconism of the Seven Wise Men.' They were men not so much marked by scientific genius as by practical sagacity and foresight in the appreciation of worldly affairs,* and their peculiar purpose seems to have been to enforce the early poets in their terse homely sayings or admonitions.† 'Their appearance,' says Mr. Grote, 'forms an epoch in Grecian history, inasmuch as they are the first persons who ever acquired an Hellenic reputation grounded on mental competency apart from poetical genius or effect.'

The sayings of these wise men were not addressed to inquiring minds, and were not therefore intended to evoke discussion. It is from this time, however, that we note the growth of the critical faculty. It was chiefly manifest in Athens, and it was due in chief part to the establishment of popular government and popular judicature. A certain power of speech then became necessary, not merely for those who intended to take a prominent part in politics, but also for private citizens to vindicate their rights or repel accusations in a court of justice.‡ That its development was greatest among the most enlightened sections of the Grecian name, and smallest among the more obtuse and stationary, is matter of notorious fact ; and it is not less true that the prevalence of this habit was one of the chief causes of the intellectual eminence of the nation generally.§

In abstract science we find the true scientific method established in the short period that separates Thucydides from Herodotus. Comte has said that 'prevision is the characteristic and the test of knowledge. If you can predict certain results, and they occur as you predicted, then you are assured that your knowledge is correct.' Now, Thucydides has told us in his preface that this is the method upon which his history is written : 'to record past facts as a basis for rational prevision in regard to the future.'‖ In recounting the details of the terrible

---

* Grote's ' History of Greece,' iv. 306.
† *Ibid.*, iv. 23   ‡ *Ibid.*, v. 257.
§ *Ibid.*, ii. 77.   ‖ *Ibid.*, v. 419.

plague at Athens he gives striking evidence of his
efficiency in this mode of treatment. 'His notice of the
symptoms,' Mr. Grote says, 'is such as to instruct the
medical reader of the present age, and to enable the
malady to be understood and identified.'* But the most
remarkable proof of his appreciation of scientific requirements we give in his own words, when he discloses his
reasons for making these careful observations : ' Having
myself had the distemper, and having seen others suffering under it, I will state *what it actually was*, and will
indicate in addition such other matters as will furnish any
man who lays them to heart with knowledge and the
means of calculation beforehand, in case the same mis-
fortune should ever again occur.'†

The art of prose composition appears to have come
with the first approach of scientific thought, being used
as a means of recording information. ' Neither the large
mass of geographical matter,' says Mr. Grote, ' contained
in the " Periegesis " of Hekatæus, nor the first map
prepared by his contemporary, Anaximander, could have
been presented to the world without the previous labours
of unpretending prose writers, who set down the mere
results of their own experience.'‡ But prose composition
was not in general use until after the Persian War, when
the requirements of public speaking created a class of
rhetorical teachers, while the gradual spread of physical
philosophy widened the range of instruction ; so that, for
speech or for writing, it occupied a larger and larger
share of the attention of men, and was gradually wrought
up to high perfection, such as we see for the first time in
Herodotus.'§

In matters of art the growth of this habit of criticism is
no less apparent. In earlier times the hand of the artist
had been restrained by religious and time-sanctioned
prejudices, and accordingly, in the execution of his work,
he had 'adhered with strict exactness to the consecrated
type of each particular god.' It was in statues of men,
especially in those of the victors at Olympia and other
sacred games, that genuine ideas of beauty were first
aimed at and in part attained, from whence they pass

---

* Grote's 'History of Greece,' v. 420.
† *Ibid.*   ‡ *Ibid.*, iv. 25.
§ *Ibid.*   ‖ *Ibid.*

afterwards to the statues of the gods.* But the emancipation of Greek art was due to those great sculptors and architects who belonged to that period of expanding and stimulating Athenian democracy, which likewise called forth creative genius in oratory, dramatic poetry, and in philosophical speculation.†

Mr. Grote observes that Grecian literature passed into the rhetoric, dialectics, and ethical speculation through the intermediate stage of tragedy. He remarks the wider intellectual range that the growth of tragedy evinced, and the mental progress it betokened as compared with the lyric and gnomic poetry, or with the Seven Wise Men and their authoritative aphorisms.‡ 'The great innovation of the dramatists,' says the historian, 'consisted in the rhetorical, the dialectical, and the ethical spirit which they breathed into their poetry. Of all this the undeveloped germ doubtless existed in the previous epic, lyric, and gnomic composition ; but the drama stood distinguished from all three by bringing it out into conspicuous amplitude, and making it the substantive means of effect.§ The characters of Grecian mythology reappear in the tragedies, but the relations between them are human and simple. They are exalted above the level of humanity, but only in such measure as to present a stronger claim to the hearer's admiration or pity. Whereas the earlier Grecian poetry was employed in recounting the exploits and sufferings of the heroes, in the tragedy the internal mind and motives of the doer or sufferer, on which the ethical interest fastens, are laid open with an impressive minuteness which neither the epic nor the lyric could possibly parallel.* The problems advanced by the dramatists, as Mr. Grote remarks, no doubt appealed to the ethical sentiment of the audience, but they were calculated also to awaken the spirit of reflective and critical inquiry, and that they had this effect it is necessary for us to admit in order to comprehend their connection with the period of philosophical speculation which followed.

But philosophical speculation could lead to no satisfactory results so long as the knowledge of each indi-

* Grote's ' History of Greece,' iv.. 25.
† *Ibid.*, v. 285.   ‡ *Ibid.*, viii. 139.
§ *Ibid.*, viii. 136.   ‖ *Ibid.*, viii. 122.
* *Ibid.*, viii. 137.

136466

vidual sacrificed its exactness to the religious or poetical
bent of his imagination. Accordingly, we now find the
Grecian intellect, under the stimulus of Sokra'es, engaged
in a supreme effort to revise and determine its knowledge
and methodize its reasoning process. To correct ' the
seeming and conceit of knowledge without the reality';
to create 'earnest seekers, analytical intellects, fore-
knowing and consistent agents, capable of forming conclusions for themselves and of teaching others '*— these
were the objects of the new intellectual movement.
Sokrates found that men were ever re.dy to give con-
fident opinions on the gravest questions concerning men
and society without bestowing upon them sufficient
reflection to be aware that they involved any difficulty.†
He convinced them out of their own mouths of the
difficulty they did not realize, and persuaded them that
their knowledge was .but ' ignorance mistaking itself for
knowledge.' By pointing out their fallacies in the definition of general terms, he introduced exactness into this
elementary part of the logical process. His efforts were
directed to 'learn what each separate thing really was,'
and his method as shown in the Discourses was to
distinguish and distribute things into genera or families,
thereby introducing that method of generalization that we
know as the inductive process. In Sokrates the increased
self-working of the Grecian mind manifested itself.‡ He
laid open, says Mr. Grote, all ethical and social doctrines
to the scrutiny of reason, and first awakened amongst
his countrymen that love of dialectics which never left
them — an analytical interest in the mental process of
inquiring out, verifying, proving, and expounding truth.

In attempting this brief abstract of the record
of so momentous a subject as the growth of
the Grecian intellect, we have been careful to
introduce no speculations of our own. Our
authority throughout has been the highest.
We have drawn our information from Mr.
Grote's monumental history of this nation, and,

* Grote's ' History of Greece,' viii. 257.
† *ib:d.*, viii. 242.   ‡ *Ibid.*, iv. 23.

as it will be observed, we have, to ensure accuracy, in great measure adhered to the actual text of the author. Various considerations suggest themselves in reviewing the growth of this expansive intellect; and though we are sensible of the extreme difficulty and peril of dogmatizing when endeavouring to extract meaning from materials of such complexity, yet we make bold to offer a few suggestions as calculated to show in what way the portrayal of this national mind illuminates the obscure problem of the development of the individual intellect.

In the first place, we may safely affirm that imagination was the earliest manifestation of mental activity of which we have record. Poetry and music formed the only intellectual food, and constituted, moreover, the sole moral stimulus, appealing as they did to the 'sympathy, emotion or reverence' of their hearers. The critical age was ushered in by maxims or proverbs which, though they enforced the teachings of the poets, yet presupposed the awakening judgment. The growth of reason was chiefly attributable to the tremendous power of speech that grew out of free institutions, and which produced that power of thought which has furnished the world with ideals which it has in vain endeavoured to excel. It manifested itself first by the display of its creative genius, and later by that specu-

lative moral and political philosophy, and the
didactic analysis of rhetoric and grammar,
which long survived.* Of the deliberate pursuit
of natural science we hear little during the
brilliant period of Grecian history, and we may
take it that scientific culture, as we understand
it to-day, was not found necessary to the pro-
duction of the intellect of this nation. That
they had acquired a large power of observation
there can be no doubt; but that they attempted
in any degree, until much later, to give systematic classification and exactness to their knowledge, we have no reason to think.

If this brief historical retrospect is now considered with regard to its bearing on the individual, the following, we believe, are amongst
some of the most obvious lessons to be derived
from it :

1. What appetite is to infancy, feeling is to
childhood and youth, the prevailing impulse
and paramount sanction.

2. Since the Grecian intellect did not attain
its phenomenal splendour until the formation
of the power of judgment, it should obviously
be our endeavour, though we must accept the
period of feeling as a necessary condition of
early years, to neutralize and regulate its control by the cultivation of the reasoning faculty.

3. In the Seven Wise Men and their un-
criticised maxims we find Reason, through the

* Grote's ' History of Greece,' v. 260.

medium of popular precepts, operating as a guide to conduct during *the age of feeling*, and operating as such, not by force of the evidence upon which the precepts were built, but by virtue of the weight of the authority by whom they were proclaimed. In the case of the individual we have a parallel to this in the acquiescence of the child in the knowledge of the adult—a parallel that, moreover, affords a curious justification of the method of the adult in obtaining such acquiescence by force of authority, rather than by an appeal to intellect.

4. That the class of intellect of which the Greek, amongst communities, is the most brilliant example, is first manifested by fertility of imagination.

5. That the memory throughout the early period is most retentive. It is matter of common knowledge that the ancient poems were handed down by oral transmission. May we not from this conclude that the memory attained and preserved its power by virtue of the use to which it was put, and that the exercise of the faculty is therefore the natural means of cultivating it?

6. That the practice of speech is a powerful aid in developing the mental faculties.

7. That the origin and growth of reading and writing are intimately associated with the process of observation.

8. From the period of infancy the mind

employed in storing up observations, but the introduction of any general method and arrangement into its accumulations is deferred until later.

9. That in art strict conformity is imposed until the mind is fitted to earn its own emancipation. This is significant, as suggesting a natural provision for preserving the mind, during the cultivation of the faculty of judgment, from an extravagant expression of feeling.

CHAPTER III.

THE DEVELOPMENT OF THE POWERS.

' Most people's minds are too like a child's garden, where the flowers are planted without the roots.'—REV J. JOYCE.

                                  ' Knowledge dwells
In heads replete with thoughts of other men :
Wisdom, in minds attentive to their own.
Knowledge—a rude unprofitable mass,
The mere materials with which wisdom builds—
Till smoothed, and squared, and fitted to its place,
Does but encumber, whom it seems t'enrich.
Knowledge is proud, that it has learned so much ;
Wisdom is humble, that it knows no more.
Books are not seldom talismans and spells,
By which the magic arts of shrewder wits
Hold an unthinking multitude enthralled.
Some to the *fascination of a name*
Surrender judgment hoodwinked.   Some the *style*
Infatuates ; and, through labyrinths and wilds
Of error, leads them by a tune entranced.
While sloth seduces more, too weak to bear
The insupportable fatigue of thought ;
And swallowing therefore, without pause or choice,
The total grist unsifted, husks and all.
But trees, and rivulets, and haunts of deer,
And sheep walks, populous with bleating lambs,
And groves, in which the primrose ere her time
Peeps through the moss, that clothes the hawthorn root,
Deceive no student.   Wisdom there, and truth,
Not shy as in the world, and to be won

By slow solicitation, seize at once
The roving thought, and fix it on themselves.'
COWPER.

' Reading maketh a full man ; conference a ready man ;
and writing an exact man.'—BACON.

' We can never reckon, then, among philosophic souls
that which is forgetful ; but we shall, on the other hand,
require it to have a good memory.'—PLATO.

MUCH has been said regarding the development
of the mental faculties, but it must not be
thought, because a distinct name is given to
each, that they therefore represent separate
divisions of the mind, and that a distinct
activity operates in each. The memory, for
instance, is not a section of the mind, but one
manifestation among many of an activity that
is one and indivisible.

To deal with those faculties and habits of mind
with which education is most nearly concerned:

Imagination.—Imagination has been thus
defined : ' *Memory* retains and recalls the past
in the form which it assumed when it was
previously before the mind. *Imagination* brings
up the past in new shapes and combinations.
Both of them are reflective of objects ; but the
one may be compared to the mirror which
reflects whatever has been before it in its proper
form and colour ; the other may be likened to
the kaleidoscope, which reflects what is before
it in an infinite variety of new forms and dis-
positions.'*

* M'Cosh's ' Typical Forms.'

Let us try to determine what part the imagination plays in the mental process. To take an elementary instance: A little child is told some story of a lion, and, as the result, his rest is disturbed by vague fears. He is then taken to see a living lion, and from that time his alarm disappears. The child now *knows* what a lion is, and the vagaries of the imagination have no longer any terror for him. The lion may, however, still constitute food for his imagination, but it will always be a lion. He may conceive one greater in size or in strength than the one he has seen, but, subject to his power to be voluntarily deceived, a story of a lion will never again picture to his mind that indefinable monster that was wont to give rise to such distress of mind.

It is thought by many that at this imaginative age the species of fiction we now call fairy-stories affords a desirable mental stimulant. We must admit that our observations in the previous chapter appear at first sight to lend some countenance to this view. It might no doubt be contended that, because in the infancy of nations the only culture passed on from generation to generation was derived from myths and legends, therefore fiction of this kind is a suitable mental food for the young child. But we must remember that, in reviewing the mental history of nations, it is of the first importance to discriminate between what has produced

good and what has produced ill results;
between what has promoted and what has
retarded the growth of character and intellect.
Now, in dealing with myths and fairy-stories,
we shall find that they are capable of division
into two distinct classes. In our illustration
of the story of the lion, we distinguished
between the effect produced by the idea of the
animal on the mind of the child before and
after he had seen it. These two conditions of
mind faithfully represent the generative causes
of the two classes of mythical creations: the
monstrous, and the natural and ideal. The
state of mind that gives birth to the unnatural
conceptions of primitive man and the child is
a negative quality. It is a condition of mind
arising from ignorance, and in fact presents
nothing more nor less than the mere phenomena
of the intellect before it has come under the
influence of experience. To attempt to increase
or prolong this condition is manifestly sub-
versive of even the least ambitious aims of
education. We take it as in no way contributing to the healthy growth of the national
mind, and therefore undesirable as an agent in
the development of the individual mind. With
myths and stories that interpret Nature, and
with those that depict the supernatural, provided their character in this respect lay in the
mere enlargement or idealization of the natural,
it is otherwise. Both for their intrinsic worth

and their power to counteract the natural habit of the infant mind to paint the monstrous, they are needful as an aid to early culture. We shall therefore be wise to follow the recommendation of Plato that care must be taken that the fiction employed in mental training is beneficial, and not mischievous.

But it is urged, chiefly from the example afforded by ancient Greece, that the fruitfulness of the national intellect in matters of art is proportionate to the abundance of the national myths. 'We must not omit,' says the historian Grote, 'the incalculable importance of the myths as stimulants to the imagination of the Grecian artists in sculpture, in painting, in carving, and in architecture. From the divine and heroic legends and personages were borrowed those paintings, statues and reliefs which rendered the temples, porticos and public buildings at Athens and elsewhere objects of surpassing admiration. Such visible reproduction contributed again to fix the types of the gods and heroes familiarly and indelibly on the public mind. The figures delineated on cups and vases, as well as on the walls of private houses, were chiefly drawn from the same source, the myths being the great storehouse of artistic scenes and composition.'

But the inspiration of Greek art did not arise from its traditional centaurs, satyrs, tritons, and other monsters, but from its idealized men and women, all natural, but perfected as types of beauty, strength and valour. In the Barberini Juno, the Belvedere Apollo, the Mars Ludovisi, the Venus of Milo, the Giustiniani Vesta, the Sleeping Ariadne, the Farnese Hercules, in Niobe—in one and all of these* the monstrous has no part, and it is upon these, and upon such as these, that the fame of Greek sculpture rests.

We must be careful, moreover, in reading our children fairy-stories, how we allow our own incredulity to infect them. It is of capital importance to bear in mind that in the infancy of nations their myths were believed, and it was only in the succeeding age of criticism that their literal truth was called in question. Moreover, it is not well to impart a habit of doubt to our children. The result may be somewhat unexpected. For we have it on the authority of Professor Sully that children graft the ideas introduced by their religious instruction on to those of fairy-lore, with some not altogether surprising confusion of the two.

If knowledge, then, be necessary for the healthy exercise of the imagination, how important a part does the acquisition of knowledge play in the cultivation of the faculty, the more so as we come to recognise that 'the office of our thought is to develop, to combine,

* We adopt the general opinion that the works of the so-called Roman school were executed by Greek artists.

and to derive, rather than to create'!* 'The
imagination of a painter or sculptor,' says Mr.
Fairholt,† 'is the fruit of genius cultivated by
study ; to depict images under the most beautiful forms, he must have that knowledge of
contour of forms which is acquired by the
practice of design ; to imagine the figures acting
in conformity with the subject, he must have
observed with meditation the movements of
man under the different actions of which he
is susceptible ; to depict the proper expression
he must have studied the effects of the affec-
tions of the mind upon the body ; to represent
the lights and colours, he must know the effects
of light upon the body, according to its position, substance or colour, as proper to each ;
and, above all, have received from Nature that
aptitude to see well and to render well those
things which constitute the genius of the
sculptor and painter.'

But we must be careful to distinguish between
the mere accumulation of ideas calculated to
feed the imagination and the active employment
of the faculty itself. The imagination actively
engaged, expressing itself in works of art, receives its vitality from the play of the emotions,
and the play of the emotions should not be
encouraged at a time when the mind should be
employed in those severer studies whose end is

* Tylor's 'Primitive Culture.'
† 'Dictionary of Terms in Art.'

4

the strengthening of the faculty of judgment.
'Nothing is more dangerous,' says Jean Paul
Richter, 'either for art or heart than the premature expression of feeling ; many a poetic
genius has been fatally chilled by delicious
draughts of Hippocrene in the warm season
of youth.　The feelings of the poet should be
closely and coolly covered, and the hardest
and driest sciences should retard the bursting
blossoms till the due spring time.'

But this introduces us to the subject of art
training, and as we purpose dealing with this
in a subsequent chapter, we must reserve until
then any further observations we may have to
make on this head.

Observation.—Observation is the direct perception of objects as they appear to us through
the medium of the senses.　It furnishes us
with the raw material from which to fashion
future knowledge.　'To observe,' says Lavater,
' is to be attentive, so as to fix the mind on a
particular object, which it selects, or may select,
for consideration, from a number of surround-
ing objects.　To be attentive is to consider
some one particular object, exclusively of all
others, and to analyze, consequently, to distinguish, its peculiarities.　To observe, to be
attentive, to distinguish what is similar, what
dissimilar, to discover proportion and dispro-
portion, is the office of the understanding.'

Can this faculty be enlarged by training ?

It is well known that the child sees more in the aggregate and in detail than the adult. But if this faculty is strong in the child and weak in the adult, is it necessarily so? Why should the power become so unproductive in the man, unless it is our neglect to cultivate it? It has until lately been customary to discourage the teaching of observation; but how, except by cultivation, can we give the individual the *habit* of observation that the power may *not be lost* to him when he grows up?

But if we cannot hope to carry this power into our later years, how important does it become that we should utilize it while it lasts! The impressions we receive in childhood are the most lasting; persons of advanced years can recall the most trivial occurrences of their youth, when the events of middle life have faded from their memory. Childhood is peculiarly the period of accumulation and cognizance of facts, and things acquired during that period are the least likely to be forgotten. That being the case, should we not encourage this capacity for observation in the child? The natural instinct of curiosity, that at this age is so active, is a safe guide to the child's mental requirements; and to refuse nutriment then is to run the risk of promoting intellectual atrophy in riper years.

We say 'teach observation.' We had rather say 'direct observation.' For the child will not require teaching in the ordinary sense. But as he is not master of his own movements, and is no judge of the value of the impressions his mind receives, it is desirable that those who are consulting his welfare should afford him full opportunity for storing up ideas, and particularly those ideas that are likely to prove most beneficial and useful to him in the future.

It is one thing to receive a general impression of an object, and another to detect its particular characteristics. The habit of observation will have been encouraged to no purpose if correctness of eye be not attained. Various artificial methods are suggested as a means of implanting this power; but none, we believe, show better results than requiring the child, in the first place, to describe in words the object that has caught its notice, and, in the second place, to draw it. Both processes carry the young eyes from the appreciation of the general effect to an examination of the detail. In the French lycées drawing is a compulsory subject, and proceeds through a course of freehand up to model drawing. ' I should make it necessary,' says Huxley, ' for every boy, for a longer or shorter period, to learn to draw. . . . In my judgment, there is no mode of exercising the faculty of observation and the faculty of accurate reproduction of that which is observed, no discipline which so readily tests error in these matters, as drawing properly taught.'

But it must be taught as a science rather than as an art, and must not be pushed too far, lest it come to engross too much of the pupil's mind, and so impede the due development of his faculties as a whole. And this is even more likely to be the result if, on completing his course of drawing, he experience the subtle allurements of colour.

Concentration, or Attention.—'When we see, hear, or think of anything, and feel a desire to know more of it, we keep the mind fixed upon the object; this effort of the mind, produced by the desire of knowledge, is called attention.'*

Concentration is mental effort prolonged by the dominion of the will. Nothing is more difficult to acquire. In the young child it does not exist, and it would be detrimental to force it. Our little ones come into a world full of delightful and marvellous things. Their interest is aroused at every turn. They have no time, even had they capacity, for prolonged attention; on every hand new sources of wonderment bear along the awakening mind. The field of experience is day by day enlarged, and the knowledge of external things builds up the young intelligence.

The general knowledge that is thus acquired will in due time give birth to powers of penetration and inquiry, the call for deeper and more

* Taylor's 'Elements of Thought.'

special knowledge. Then the power of concentration will grow, and then it should be cultivated.

There is no greater stimulus to concentration than the interest afforded by the subject. The mind gives its attention, naturally, to what interests it, but when the interest wanes, attention is maintained only by force of discipline. ' Locke is justified,' remarks Canon Daniel, ' in saying that persons may remember well one class of things and not another. The reason is obvious. We remember what we attend to, and our attention is mainly dependent on our interest. As our interest in different things varies, our power of remembering them varies also.'

It should be noted that all studies require concentration, but the pursuit of mathematics demands the most sustained attention, and hence is perhaps the most helpful in developing this power. ' There is no defect in the faculties intellectual,' says Lord Bacon, ' but seemeth to have a proper cure contained in some studies; as, for example, if a child be bird-witted, that is, hath not the faculty of attention, the mathematics giveth a remedy thereunto, for in them, if the wit be caught away for a moment, one is new to begin.'

Classification.—This is ' the sorting of a multitude of things into parcels, for the sake of knowing them better, and remembering them more easily. When we attempt to classify a multitude of things, we first observe some respects in which they differ from each other; for we could not classify things that are entirely alike, as, for instance, a bushel of peas; we, then, separate things that are not alike, and bring together things that are similar.'*

Classification is the principle of order in knowledge. When we consider the vast multitude of facts that are comprised in any one branch of knowledge, the need for arrangement becomes apparent. Of what use would a library be, the books of which were placed on the shelves irrespective of their subject-matter, or, worse still, thrown in a heap on the floor? But arrangement in knowledge is of still greater importance. It is by the classification of facts that principles are arrived at; and principles or rules are the lines upon which the human mind operates, and arrives at new facts and conclusions.

'All *science*,' says Mr. Lewes,† 'consists in the *co-ordination* of facts. If our different observations were entirely *isolated*, there would be no science. We may even say that, in so far as the different phenomena will permit, *science* is essentially *destined to dispense with all direct observation*, by allowing us to *deduce the*

* Taylor's ' Elements of Thought.'
† Comte's ' Philosophie Positive,' by G. H. Lewes.

greatest possible number of results from the smallest possible number of immediate data.'

'And so,' says Froebel,* 'since education has largely to do with inducing the right acquirement of knowledge and the right use of knowledge, the task of the educator must largely consist in bringing out, and making clear, and maintaining, the connectedness of facts and things.'

Memory.—The memory is that faculty which enables us to treasure up and preserve for future use the knowledge we acquire—a faculty which is obviously the great foundation of all intellectual improvement, and without which no advantage could be derived from the most enlarged experience. This faculty implies two things: a capacity of retaining knowledge, and a power of recalling it to our thoughts when we have occasion to apply it to use.'†

The power of memory has a close connection with the habit of classification. 'Recollection,' says Jean Paul Richter, 'like every other mental power, can only work according to the laws of association. . . . Read a volume of history to a boy, and compare the copious abstract he can furnish of that with the miserable remnant he could collect from a page of Humboldt's Mexican Words which you had read aloud to him.'

* H. C. Bowen's ' Froebel.'
† Stewart's ' Philosophy of the Human Mind.'

Quintilian, whose memorable work was said by John Stuart Mill to contain all that was best in the educational doctrines of the ancients, has given some valuable advice regarding the cultivation of the memory.

In an earlier part of his work he observes that the chief symptom of ability in children is memory, of which the excellence is twofold : to receive with ease and retain with fidelity. He then insists on the necessity for stimulating the faculty, reminding us that all knowledge depends on memory ; and that we shall be taught to no purpose if whatever we hear escapes us, and concludes with some practical recommendations, from which we extract the following :

' If anyone ask me, however, what is the only and great art of memory, I shall say that it is *exercise* and *labour*. To learn much by heart, to meditate much, and, if possible, daily, are the most efficacious of all methods. Nothing is so much strengthened by practice, or weakened by neglect, as memory. Let children, therefore, as I directed, learn as much as possible by heart at the earliest possible age.'

' For fixing in the memory what we have written, and for retaining in it what we meditate, the most efficacious, and almost the only, means (except exercise, which is the most powerful of all) are division and arrangement. He who makes a judicious division of his subject will never err in the order of particulars.'

Taste.—'Taste is the settled habit of discerning faults and excellencies in a moment—the mind's independent expression of approval or aversion. It is that faculty by which we discover and enjoy the beautiful, the picturesque, and the sublime in literature, art, and nature.'

For the cultivation of taste in poetry, art, or music, it is not necessary to be either a poet, an artist, or a musician; but it is necessary to have an acquaintance with poetry, art, or music. As in the case of the imagination, knowledge is the first requirement. A high standard of excellence is necessary, and this can be acquired only by an appreciation of the best that has been done, and a discernment of the possibilities of higher achievements. It is grounded upon knowledge, but a knowledge of the best.

Reasoning.—'Reasoning is that operation of the mind through which it forms one judgment from many others; as when, for instance, having judged that true virtue ought to be referred to God, and that the virtue of the heathens was not referred to Him, we thence conclude that the virtue of the heathens was not true virtue.'*

The pursuit of every art is preceded or is accompanied by the knowledge of its correlative science. The art of medicine, to wit, succeeds the study of anatomy and physiology. The art of clockmaking is not acquired

* 'Port Royal Logic.'

without a co-extensive knowledge of the principles of construction. The art of reasoning,
also, is not without its corresponding science,
and that science is knowledge itself. But wide
knowledge is not identical with skill in reasoning, any more than an intimate knowledge of
anatomy and physiology or the principles of
construction of a clock is identical with skill in
surgery or clockmaking.

While much time and much money are spent
in learning the arts of reading and writing, little
or no attention is given to the art of reasoning. Since we must reason if we are to get
through life at all creditably, is it not desirable
that we should lose no time in learning to
reason well? We devote many hours of our
children's lives to achieve skill in pianoforte-
playing, but we give none to the acquisition
of power in reasoning. If it is desirable that
we should fit ourselves early to pursue an art
which is voluntary, how much more desirable
is it that we should fit ourselves early to pursue
an art that is compulsory!

There is a difference of opinion as to the
best mode of cultivating the faculty of reason.
The science of logic, John Stuart Mill contends,
'is the great disperser of hazy and confused
thinking; it clears up the fogs which hide from
us our own ignorance, and make us believe
that we understand a subject when we do not.'
But Dr. Whewell considers that to cultivate

logic as an art is like learning horsemanship by book, and another author holds that the principles of logic are best acquired by the study of those authors who reason the best. Locke averred that he knew of no case of skill in reasoning being acquired by the study of the rules of logic ; but he would no doubt have been in accord with John Stuart Mill when he observed that ' if the practice of thinking is not improved by rules, I venture to say it is the only difficult thing done by human beings that is not so.'

Dr. Whewell, however, is not absolute in his opinion, for elsewhere he writes : ' Let us suppose it established, then, that it is a proper object of education to develop and cultivate the reasoning faculty. The question then arises, By what means can this be done ? What is the best instrument for educating men in reasoning ? There are two principal means which have been used for this purpose in our Universities : the study of mathematics and the study of logic. These may be considered respectively as the teaching of reasoning by practice and by rule.'

The faculty of reason is, we believe, best matured by the actual practice of the art of reasoning. Useful, as it no doubt is, to be familiar with the best models of reasoning, it is a familiarity that produces increase of knowledge, not increase of skill. We may be acquainted with the best works of art, and our knowledge may extend to both their technical and artistic merits; but this acquaintance will not enable us to create works of equal worth. It is only practice that could do that, and even practice can do it only on certain conditions.

The practice of the art of reasoning is active thought precipitated, as it were, in the form of speech or writing. Conference, says Lord Bacon, maketh a ready man, and writing an exact man. 'Certain it is,' he observes, 'that whosoever hath his mind fraught with many thoughts, his wits and understanding do clarify and break up in the communicating and discussing with another; he tosseth his thoughts more easily; he marshalleth them more orderly; he seeth how they look when they are turned into words; finally, he waxeth wiser than himself, and that more by an hour's discourse than by a day's meditation. . . . In a word, a man were better relate himself to a statue or picture than to suffer his thoughts to pass in smother.'

It is not until conversation is regulated by established method, and becomes the art of disputation, or debating, that its full disciplinary value is perceived. It then becomes a great power in quickening our perceptive and reflective faculties and in facilitating the adequate and spontaneous expression of our thoughts.

'Disputation,' says Sir William Hamilton, 'is, in a certain sort, the condition of all im-

provement.' It brings out, he contends, 'the
most important intellectual virtues: presence
of mind, dominion over our faculties, prompti-
tude of recollection and thought, and withal,
though animating emulation, a perfect command
of temper. It stimulates also to a more attentive and profounder study of the matters to
be thus discussed; it more deeply impresses
the facts and doctrines taught upon the mind;
and, finally, what is of peculiar importance,
and peculiarly accomplished by rightly-regulated
disputation, it checks all tendency towards
irrelevancy, and disorder in statement, by
astricting the disputants to a pertinent and
precise and logically-predetermined order in the
evolution of their reasonings.'

De Quincey, avowedly a taciturn man, en-
larges upon the new light that awaited him
when he came to realize the educational value
of the practice of speaking. Yet the culture
of this faculty, in our modern school systems,
is almost entirely disregarded. And what is
more indispensable to the active man than a
fluent and well-regulated power of speech?
Oratory has been termed the art of persuasion,
and, though it is not needful to the majority
of men, every man is at a disadvantage who
cannot express himself with distinctness, fluency,
and point. In most cases these qualities may be
attained by training and practice. 'If, therefore,'
says Quintilian, 'we have received from the

gods nothing more valuable than speech, what can we consider more deserving of cultivation and exercise ? or in what can we more strongly desire to be superior to other men than in that by which man himself is superior to other animals, especially as in no kind of exertion does labour more plentifully bring its reward ?'

It is curious, but it is no less true, that the man who is logical in speech is not necessarily logical on paper, and the reverse is no less a fact. The criticism that is brought to bear on the logic of the speaker is less penetrating than the criticism that reviews the work of the writer. The argument of the speaker is no sooner delivered than the attention of the listener is carried on to new matter; but the reasoning of the author is crystallized in permanent type over which the critic can pore with relentless concentration. Therefore, by reason of the greater demands upon the author, we conclude that the practice of writing can develop greater logical accuracy than the practice of speaking. The highest achievement is reached when skill is attained in both.

CHAPTER IV.

THE CLASSICS.

'Then farewell, Horace ; whom I hated so !'

BYRON.

' Latin and Greek are a great and splendid ornament, which commonly, however, is too dearly bought.'—MONTAIGNE.

' Happy were the Latins, who needed only to learn Greek, and that not by school teaching, but by intercourse with living Greeks. Happier still were the Greeks, who, so soon as they could read and write their mother-tongue, might pass at once to the liberal arts and the pursuit of wisdom. For us, who must spend many years in learning foreign languages, the entrance into the gates of Philosophy is made much more difficult. For to understand Latin and Greek is not learning itself, but the entrance - hall and ante-chamber of learning.'—MELANCTHON.

' The antique symmetry was the one thing wanting to me.'—LEONARDO DA VINCI.

' For I cannot help thinking that classical literature, in spite of its enormous prestige, has very little attraction for the mass even of cultivated persons at the present day. I wish statistics could be obtained of the amount of Latin and Greek read in any year (except for professional purposes), even by those who have gone through a complete classical curriculum. From the information that I have been able privately to obtain, I incline to think that such statistics, when compared with the fervent admiration with which we all still speak of the Classics upon every opportunity, would be found rather startling.' —SIDGWICK.

By the Classics we understand the study of the Greek and Latin languages and literatures.

For many years a controversy has existed with regard to the advantage or the comparative advantage of devoting so large a portion of school hours to this study; and of late, as the claims of science have become more fully recognised, the question has become a more vital one. The same battle is being fought out on the Continent, and the present trend of opinion, Matthew Arnold tells us, in speaking of France and Germany, is 'a growing disbelief in Greek and Latin—at any rate, as at present taught— and a growing disposition to make modern languages and the natural sciences take their place.'

The objections to the study itself, or the methods adopted in its pursuit, are not new, for Milton, in his letter on Education, thus delivers himself on the subject : 'We do amiss to spend seven or eight years merely in scraping together so much miserable Latin and Greek as might be learned otherwise easily and delightfully in one year. And that which casts our proficiency therein so much behind is our time lost in too oft idle vacancies given both to schools and universities; partly on a preposterous exaction, forcing the empty wits of children to compose themes, verses and orations, which are the acts of ripest judgment, and the final work of a head filled by long reading and observing with elegant maxims and copious invention.   These are not matters to be wrung from poor striplings like blood out of the nose.'

Milton's indictment is preferred in our own day, if not with more truth, yet perhaps with greater vehemence.   ' Is it not madness,' writes Jean Paul Richter in the ' Levana,' ' to think it even possible that a boy of fourteen or sixteen, however *great* his abilities, can comprehend the harmony of poetry and deep thought contained in one of Plato's discourses, or the worldly persiflage of Horace's satires, when the genius itself has not conducted the men I name to the pure cold heights of antiquity until long after the fiery season of youth ? . . . Merely verbal difficulties may be overcome by teaching and industry, but mental difficulties only by the maturity of thought which comes with years.'

In the Middle Ages Latin was the language of the cultured classes, and the literature of Europe in those times was for the greater part a litera-ture written in that tongue.   It was made the groundwork of education, says Mr. Parker, ' not for the beauty of its classical literature, not because the study of a dead language was the best mental gymnastic or the only means of acquiring a masterly freedom in the use of living tongues, but because it was the language of educated men throughout Western Europe, employed for public business, literature, philosophy and science; above all, in God's providence, essential to the unity, and therefore enforced by the authority, of the Western Church.'*

But this inducement to the study of Latin has ceased to have weight. Beyond its employment in scientific nomenclature, and its very limited service to the legal practitioner, the study of this language is advocated on entirely new grounds. Moreover, Greek, viewed with regard to its practical utility, lays no class of the community under obligations, except, it may be, the clergy.

If, then, the classical tongues have ceased to be *necessary* to us to-day, it will be well for us to consider upon what grounds the study of them is held to be *desirable.* It is not to be denied that few scholars push their classical studies to that point that gives them access to the ancient literatures. Nor is it seriously maintained that the mere knowledge of the languages has any intrinsic value. Yet we are told that the Classics are the backbone of a liberal education. The average public schoolboy presents all the intellectual limitations that this deliberate mental starvation involves. His

* ' Essays on a Liberal Education.'

prejudices are profound, proportionate, in fact,
to his ignorance on all matters that lie without
the bounds of his narrow curriculum.   His
knowledge of the English vocabulary does not
greatly exceed the equivalent in the vernacular
of those ideas that a dead language can supply.
' The half-technical, the philosophical language,'
says Mr. Sidgwick, 'which thoughtful men
habitually use in dealing with abstract subjects,'
he is in danger of never learning. 'Of some
of these terms,' says the same author, 'such a
boy may pick up a loose and vague comprehension from ordinary conversation, novels and
newspapers, but he will generally retain sufficient ignorance of them to make the perusal of
all difficult and profound works more weary and
distasteful than their subject matter alone could
make them.'

Nor can it be contended that the study of
ancient languages is a literary culture.   Where
is the boy who, leaving school at seventeen,
with his classical studies not half finished, does
not from that day instinctively identify his
Cicero, his Xenophon and his Euripides with
grammatical problems and linguistic difficulties?
The consequence is, says Mr. Sidgwick, 'that
half the undergraduates, and a large proportion
of the boys at all (except perhaps one or two) of
our public schools, if they have received a literary
education at all, have got it for themselves;
the fragments of Greek and Latin that they

have struggled through have not given it them.
If so many of our most expensively-educated
youths regard athletic sports as the one con-
ceivable mode of enjoying leisure ; if so many
professional persons confine their extra-pro-
fessional reading to the newspapers and novels ;
if the middle-class Englishman (as he is continually told) is narrow, unrefined, conventional,
ignorant of what is really good and really evil
in human life ; if (as an uncompromising
writer says) he is " the tool of bigotry, the echo
of stereotyped opinions, the victim of class
prejudices, the great stumbling-block in the
way of a general diffusion of higher cultivation
in this country," it is not because these persons
have had a literary education which their " in-
vincible brutality " has rendered inefficacious ;
it is because the education has not been (to
them) literary : their minds have been merely
put through various unmeaning linguistic exercises.'

After this heavy indictment of the customary
mode of study of the classical tongues and its
results, let us proceed to consider what the
advocates of the system have to say in support
of it.

The study of the Greek and Latin languages,
and of the authors who have written in them,
is, says Dr. Donaldson, the particular form
which our grammatical teaching has assumed.
For, he says later, 'the ultimate object of

classical training is to give the many a habit of methodically arranging their thoughts.'

This habit, it would seem, is acquired by the structural analysis of the languages that accompanies and forms part of our study of them. In illustration of this, the Rev. Thomas Hughes observes: 'I may be allowed the passing remark, which is familiar to every judge of a classical education, that the disciplinary value of literary studies reaches here its highest degree of mental exercise; and that the two classical tongues, Latin and Greek, are altogether eminent as supplying materials for this exercise in their own native structure; which in the Latin is an architectural build, characteristic of the reasoning human mind; and in the Greek is a subtle delicacy of conception and tracery, reflecting the art, the grace and versatility of Athens and the Ionian Isles.'

But the chief, and perhaps the only additional, advantage afforded the classical student, who has not acquired facility in reading in the original, is the knowledge he gains of the structure, derivation, and pronunciation of his own tongue, together with a better and more precise acquaintance with the meaning of the words and phrases that compose it. The practice of translating, and more especially from the classical languages, is particularly conducive to this accuracy. 'As we seldom think of asking the meaning of what we see every day,' remarks John Stuart Mill, 'so, when our ears are used to the sound of a word or a phrase, we do not suspect that it conveys no clear idea to our minds, and that we should have the utmost difficulty in defining it, or expressing in any other words what we think we understand by it. Now, it is obvious in what manner this bad habit tends to be corrected by the practice of translating with accuracy from one language to another, and hunting out the meanings expressed in a vocabulary with which we have not grown familiar by early and constant use.'

But are not these benefits alleged to be derived from classical tuition small when we take into account the price paid for them, viz., the almost entire appropriation of the scholar's most receptive years? It is not claimed that there is any intrinsic advantage in the mere knowledge of a dead language. The study of it is a means to an end, but to an end it may not be the best means to attain.

It is admitted that in so far as language can furnish the mental exercise afforded by science it is preferable, by reason of the greater convenience that naturally attends tuition from a book, than tuition by means of natural objects and scientific apparatus. We acknowledge, too, that the science of language confers the peculiar benefits attaching to a science of its class. But it does not comprise in its result the full effect

that is produced by scientific education.   It fails
to exercise that important function of the mind,
the reasoning from 'cause to effect'—a habit
that can only be acquired by the cultivation of
the scientific method that illustrates it.

While we give full credit to the value of
language as a disciplinary study, we must
again draw attention to the fact that, in the
case of the classical tongues at least, it has no
advantage in itself. 'We are guilty of something
like a platitude,' says Herbert Spencer, 'when we
say that throughout his after-career a boy in nine
cases out of ten applies his Latin and Greek to
no practical purposes. The remark is trite, that
in his shop or his office, in managing his estate
or family, in playing his part as director of a
bank or a railway, he is very little aided by this
knowledge he took so many years to acquire.'
With science, on the other hand, the education
that trains and develops the faculties at the
same time stores the mind with knowledge and
gives it power that will be of practical service
to the student throughout life. With a mind
trained in scientific method, with a general
knowledge of scientific facts, the student will
carry into his future career the intelligence
required to build up the successful man of
action. Whether as a manufacturer, doctor,
or engineer, he will bring a mind trained to
grasp with facility the knowledge and principles
special to each of these pursuits ; and he will

escape the indignity, so well described by Lord
Houghton, in speaking of the English gentle-
man, of being 'a landed proprietor without a
notion of agriculture—a coal-owner without
an inkling of geology—a sportsman without
curiosity in natural history.'

It cannot be denied that the study of the
classical tongues is an important aid in the
study of English; but is not its importance
unduly exaggerated, and is it altogether an un-
mixed benefit? As Mr. Sidgwick inquires, if
etymology is necessary for the understanding
of English, then why are the Classics taught and
Early English neglected? As we have elsewhere shown, a knowledge of the principles
of structure of a language is not necessary
to ensure grammatical expression. Even less
so is a knowledge of its derivation and
lineage. On the contrary, derivative meanings
are not generally the same as modern meanings; and he who persistently adheres to the
former will soon be convicted of pedantry.
In English, custom is the only safe guide, and
consequently the language itself, as best spoken
to-day, is the only safe study.

The process of translating undoubtedly brings
with it the advantages attributed to it by Mr.
Mill; but if it be confined to the translation of
the ancient languages, while the student gains
in cultivating the military precision that is so
essential a feature of the Classics, he loses

by reason of the necessary absence of modern ideas in the literature; so that, in representing in English the ideas of the old world, the student derives no acquaintance with the immense vocabulary that is the peculiar product of the new.

The knowledge to be acquired by those who have mastered the classical tongues and gained access to the literatures has been the topic of many a brilliant essay. It is a subject that stirs the imagination and gives birth to fine periods. We should like to quote freely, for the sake of the pleasure of introducing to our readers some of the most brilliant passages in literature with which we are acquainted; but we must confine ourselves to our purpose, and be content with selecting merely those observations which we feel will best illustrate the subject under consideration.

The learning of the Greeks and Romans is regarded as the surest foundation upon which modern civilization can rest. In tracing back the history of nations, we can find no equivalent, in stability and permanence, to the productions of these two races. Posterity has reversed fewer of their conclusions than of those of other peoples. Their ideas have been again and again ' submitted to the crucible of human thought,' until they have come to be regarded as a ' mirror or picture of human reason in general.'*

* Steinthal.

They are the nations who have been most ' industrious after wisdom,' and they afford more than any other nation 'the experience and tradition for all kinds of learning.'

Modern civilization has been grounded upon this enduring basis. ' The cultivated mind,' obsèrves Dr. Whewell, ' up to the present day has been bound together, and each generation bound to the preceding, by living upon a common intellectual estate. . . . All the countries of lettered Europe have been one body, because the same nutriment, the literature of the ancient world, was conveyed to all, by the organization of their institutions of education. The authors of Greece and Rome, familiar to the child, admired and dwelt on by the aged, were the common language, by the possession of which each man felt himself a denizen of the civilized world.'

Of the disciplinary value of ancient thought, as expressed in the classical authors, as distinct from the disciplinary value of the languages themselves, John Stuart Mill writes : ' Human invention has never produced anything so valuable, both in the way of stimulation and discipline to the inquiring intellect, as the dialectics of the ancients, of which many of the works of Aristotle illustrate the theory, and those of Plato exhibit the practice. No modern writings come near to these, in teaching, both by precept and example, the way to investigate truth, on these subjects, so vastly important to us, which remain matters of controversy from the difficulty or impossibility of bringing them to a directly experimental test.'

Of the correctness of such an estimate, Dr. Arnold furnishes us with a remarkable illustration. In a letter to the Archbishop of Dublin he describes the mental condition of a Jew from whom he was taking lessons in Hebrew, and who, learned as he was in the writings of the Rabbis, was totally ignorant of all the literature of the West, ancient or modern. 'He was consequently,' said Dr. Arnold, 'just like a child, his mind being entirely without the habit of criticism or analysis, whether as applied to words or things ; wholly ignorant, for instance, of the analysis of language, whether grammatical or logical ; or of the analysis of a narrative of facts, according to any rules of probability external or internal. I never so felt the debt which the human race owes to Pythagoras, or whoever it was that was the first founder of Greek philosophy.'

It is a curious fact that science, antagonistic as it is to classical study, has more especially appropriated Latin as its peculiar tongue, and has thus made its preservation more than ever necessary. 'The Latin language,' De Quincey says, ' has a planetary importance ; it belongs not to this land or that land, but to all lands where the human intellect has obtained its rights and its development. It is the one

sole *lingua franca;* that is, in a catholic sense,
it is such for the whole humanized earth, and
the total family of man. It is still the common
dialect which binds together that great *impe-
rium in imperio*—the republic of letters.'

Valuable as Latin is to the republic of letters,
it is hardly less so to the republic of science,
forming as it does the common ground by means
of which inquirers of different nationality can
make themselves intelligible to one another.

But perhaps the chief advantage of a close
acquaintance with Greek and Roman traditions
is attributable to the many standards of excellence with which we are familiarized—standards
that by common consent have never since been
reached, much less surpassed. ' In purely
literary excellence, in perfection of form,' remarks John Stuart Mill, 'the pre-eminence of the
ancients is not disputed. In every department
which they attempted, and they attempted
almost all, their composition, like their sculpture, has been to the greatest modern artists
an example to be looked up to with hopeless
admiration.'

' In studying the great writers of antiquity,'
he elsewhere observes, ' we are not only learning to understand the ancient mind, but laying
in a stock of wise thought and observation, still
valuable to ourselves, and at the same time
making ourselves familiar with a number of
the most perfect and finished literary com-

positions which the human mind has provided.'

But Mr. Mill attributes even a higher value
to what he designates their ' wisdom of life '—
'the rich store,' he explains, ' of experience of
human nature and conduct, which the acute
and observing minds of those ages, aided in
their observations by the greater simplicity of
manners and life, consigned to their writings,
and most of which retains all its value.'

But why, it will be asked, cannot all this be
attained by the majority of persons through
translations ?

The answer is obvious : Much of it can.
'All knowledge,' says De Quincey, 'is translatable, and translatable without one atom of
loss.' 'For no man,' says the same author,
' will wish to study a profound philosopher but
for some previous interest in his doctrines, and
if by any means a man has obtained this, he
may pursue this study sufficiently through translations.' But where the reader seeks art and
not science, then translations are all but valueless ; for ' if the golden apparel is lost, if the
music has melted away from the thoughts, all,
in fact, is lost.'

Lord Houghton considers that franker recog-
nition should be given to the worth and use of
translations into modern languages, and maintains that they should be ' the most effective
material of school training, instead of being

prohibited and regarded as substitutes for severe study and inducements to juvenile indolence.'

Mr. Sidgwick even goes further, and gives it as his opinion that the greater part of the vivid impressions that most boys receive of the ancient world are derived from English works; and, moreover, he remarks elsewhere, some persons 'would perhaps be ashamed to confess how shallow an appreciation they had of Greek art till they read Goethe and Schiller, Lessing and Schlegel.'

In conclusion, let us venture a word of practical advice. Whatever may be the disadvantages of an exclusively literary education, they can hardly exceed the drawbacks that we believe would accompany an exclusively scientific one. There is need for both. But we maintain that though it may be desirable, for the purposes of mental discipline, to study one classical tongue, the study of two does not offer additional advantages proportionate to the extra time employed in its pursuit. In the choice of Latin or Greek, we shall be guided by one of two considerations. If our choice is determined by the value of the literature (to which, after all, we may never gain access), we shall select Greek. If, on the other hand, we seek those advantages which the defenders of a classical education contend accompany the study of the ancient languages, we shall choose Latin. In selecting the latter, our choice will be wisely exercised ; for, since few boys mature their classical studies, the only benefit accruing to them is the disciplinary one we have referred to, and this they will get in greater measure from Latin than from Greek. Moreover, in nearly all examinations, professional or otherwise, Latin is a compulsory subject, and Greek is not. Obviously this is a consideration of great importance, and one that we cannot afford to overlook.

CHAPTER V.

GRAMMAR—MODERN LANGUAGES—HISTORY.

'He that is learning to arrange his sentences with order is learning to think with accuracy and order.'— BLAIR.

'It is the evident failure to carry out the original intention of classical studies which has made it necessary to bring more prominently forward the supposed advantages of grammar.'—BOWEN.

'But living languages are so much more easily acquired by intercourse with those who use them in daily life . . . that it is really waste of time for those to whom that easier mode is attainable to labour at them with no help but that of books and masters.'—MILL.

'The only history that is of practical value is what may be called descriptive sociology. And the highest office which the historian can discharge is that of so narrating the lives of nations as to furnish materials for a comparative sociology, and for the subsequent determination of the ultimate laws to which social phenomena conform.'— SPENCER.

Grammar.—If we accept the dictum of Herbert Spencer, that the education of the child must accord both in mode and arrangement with the education of mankind, considered historically,

6

then we must put the study of grammar much later in the school course than is customary. 'The study of language,' says Mr. Bowen, 'is at the present day the only kind of study which deliberately professes to advance in a direction exactly the reverse of every other branch of human progress. In every other fruitful inquiry we ascend from phenomena to principles. In classical study alone we profess to learn principles first, and then advance to facts.'

The study of grammar is not in general necessary to ensure grammatical expression of our thoughts; not more, in fact, than our reasoning power is dependent on our mastery of works on logic. 'English grammar,' observes Sir John Lubbock, 'as it is ordinarily taught in elementary schools, seems to be of very doubtful value. Moreover, the power of speaking grammatically is more a matter of practice and tact than of tuition. I do not wish to undervalue grammar with reference to language, but would say, in the words of George Herbert:

'" Who cannot dress it well, want wit, not words." '

Both Dr. Arnold and Matthew Arnold maintain that the rationale—the explanation of the principles of language—belongs to a more advanced age of scholarship, and that what is required in the school grammar is a system of 'clear categories' which the student can understand and apply practically to his reading.

'But we, and the Germans, too,' remarks the latter, 'keep trying to put the rationale of grammar into the first grammar—the grammar that is learnt, not consulted; and the boy's mental digestion rejects the rationale, and meanwhile the fixity needed for categories to which he is promptly and precisely to refer all his cases—an effort of which his mind is perfectly capable—is sacrificed.'

Nothing is so apt to provoke an early disgust of literature, or to cramp the expansion of the individual power of expression, as the selection of standard passages of poetry or rhetoric as mere exercises in grammatical analysis or construction. Grammar is the servant, not the master, of language, and the best intellects have always so regarded it; otherwise language could have no development.

The rules of grammar become necessary, says Locke, when it is thought time to put anyone upon the care of polishing his tongue. They are necessary as a preparation for the cultivation of style—an art which De Quincey, that master of style, has defined as the 'management of language,' and of the functions of which he has given the following pregnant description. 'Style,' he says, 'has two separate functions: first, to brighten the intelligibility of a subject which is obscure to the understanding; secondly, to regenerate the normal *power* and impressiveness of a subject which has become dormant to the sensibilities. Darkness gathers upon many a theme, sometimes from previous mistreatment, but often from original perplexities investing its very nature. Upon the style it is, if we take that word in its largest sense—upon the skill and art of the developer—that these perplexities greatly depend for their illumination.'

Modern Languages.—The cultivation of the English language is the chief concern of every Englishman. Is it not strange, then, that of all school subjects it is the one that receives least attention?

Many vague and superficial statements have been made in the attempt to define and explain this study. It is not merely a knowledge of the classics of the English tongue; it is not merely a knowledge of the English grammar. It is rather a study whose aim is to acquire a wide and exact knowledge of the English vocabulary, an instinctive acquaintance with the rules of grammatical composition, and the power of elegant and appropriate expression.

'Language is the medium for expressing our thoughts.' We may have nebulous ideas on many subjects, but we are unable to communicate them to our fellow-men because we are unfamiliar with that fragment of the English vocabulary that expresses them. As our knowledge of the vocabulary expands, our field of thought and power of speech enlarge. The study of English classics is altogether in-adequate for this purpose. The fact that they are classics endows them with a certain re-verend age, in consequence of which their pages are not likely to afford acquaintance with that multitude of new ideas that is the product of our century.

The teaching of English is a fit subject for home and early training. The first efforts to build up the vast vocabulary which the cultured Englishman commands must be confined to those familiar things and their parts that are the common objects of the home and country. The habit of description should be encouraged ; and the ideas of things should not only be grasped by the child, but he should be induced to express the ideas in words. Thus the know-ledge of things and the power of language will be contemporaneously developed.

When the student proceeds to abstract ideas, he should be helped to gradually accumulate those half-technical, philosophical words that are the stock-in-trade of the thoughtful, cul-tured man. ' If,' says Mr. Sidgwick, ' English authors were read in schools so carefully that a boy was kept continually ready to explain words, paraphrase sentences, and summarize arguments ; if the prose authors chosen gradu-ally became, as the boy's mind opened, more difficult and more philosophical in their diction;

if, at the same time, in the teaching of natural science, a great part of the technical phraseology (from which the main stream of the language is being continually enriched) was thoroughly explained to him, then we might feel that, by direct and indirect teaching together, we had imparted a complete grasp of what is probably the completest instrument of thought in the world.'

When the boy's vocabulary has assumed dimensions great enough to enable him to express his ideas with facility, he should proceed to the study of grammar, and particularly of syntax, that he may learn that language ' is not capricious and arbitrary in its arrangements, but reflects the operations of the mind.'*

Facility of expression and accuracy of composition attained, the student should cultivate style; and to do this he can seek no better guides than the masters of style who have written in our tongue. Not that the style of any given writer should be imitated, for style is essentially an idiosyncrasy, the peculiar and distinct expression of the individual. But a general acquaintance with style in its various garbs will imperceptibly affect the student, and slowly and surely give to his expression the peculiar characteristic in style that is the natural product of his individuality.

* J. W. Hales.

Next to English, which language has the chief claim to the student's attention ?

'French and German,' says Huxley, 'and especially the latter language, are absolutely indispensable to those who desire full knowledge in any department of science.' And there are other reasons, for later in the same work he remarks : 'If the time given to education permits, add Latin and German. Latin, because it is the key to nearly one-half of English ; and German, because it is the key to almost all the remainder of English, and helps you to understand a race from whom most of us have sprung, and who have a character and a literature of a fateful force in the history of the world, such as probably has been allotted to those of no other people except the Jews, the Greeks, and ourselves.'

De Quincey has dealt with this subject, and has laid it down as an axiom that the act of learning a language is in itself an evil ; and he has recommended us to so frame our selection of languages that the largest possible body of literature *available for our purposes* may be put at our disposal with the least possible waste of time and mental energy.

After excluding Italian, Spanish, and Portuguese, he maintains French and German to be worthy rivals for the first place, and gives the preference to the latter, with this recommendation—that German literature, for vast compass, variety, and extent, far exceeds all others as a depository for the current accumulations of knowledge.

There can be no doubt that the choice may be exercised on other than literary considerations. For instance, for purposes of statesmanship French is indispensable ; to the musician, Italian is desirable. But these are, and will remain, special considerations. For purposes of general culture, science, business, and all antiquarian matters, it is plain that German is the language that gives the amplest returns.

In learning a foreign modern tongue, our chief object is to speak, and this accomplishment can at no time be so easily acquired as in childhood. Childhood is a period when Nature gives us peculiar facilities for learning to speak a language, and in this respect a foreign nurse, in the first six or seven years of life, will do more service than the most accomplished masters throughout the succeeding school period. But the advantages to be derived from a foreign nurse will, when her influence is withdrawn, be quickly lost, unless the child be induced to continue to exercise the power he has by this means acquired ; and for this reason it is desirable that the practice of reading the language should be grafted on the habit of speech.

What we have said of the study of English applies equally to the study of a foreign tongue.

The first requirement is a substantial vocabulary ; and grammar, which in our schools so often takes its place in order of time, should be deferred until the linguist has acquired complete facility in the expression of his ideas.

History.—It is a common objection that history, as learnt at school, is limited to the mere committing to memory of dates, and that school histories give an undue prominence to mere events. History, it is urged, should be less a record of a nation's battles and a monarch's deeds than the life of a people.

These critics seem to recommend, in the study of history, a system that is expressly deprecated in the study of grammar. Children are as little likely to understand the rationale of history as they are the rationale of grammar. They must acquire the facts, and from these they will build up the principles. And what is more necessary to a correct application of the events of history than the precise date of their occurrence ? An error in a date may occasion the cause to be mistaken for the effect, and then how misleading a study will the philosophy of history become ! Dates are as necessary for a correct interpretation of history as are the positions of the planets at any given time for the ascertainment of their orbits. We do not commit ourselves to assert that a child should learn a long string of dates by heart. We merely say that, without a due

regard to dates, he will never learn to appreciate the lessons that history can impart.

It must be remembered, too, that in learning
a date he is at the same time learning a fact,
and from the association of the two ideas the
memory will better retain both.

But the facts of history, like the facts of
science, are barren and useless until the lessons
they are capable of yielding have been drawn
from them. 'The principal and proper work
of history,' says Hobbes, 'is to instruct and
enable men, by the knowledge of actions past,
to bear themselves prudently in the present,
and providentially towards the future.' For
what is true of the individual is true of the
nation: 'whatsoever a man sows, that shall
he also reap.'

The term 'history,' however, is capable of
wider signification. The popular definition,
limiting its application to the history of communities only, is too restrictive. The process
of development, which is characteristic of
every idea and everything of which we are, or
can become, cognizant, is tantamount to saying
that they have their histories. It is generally
admitted that we can understand nations only
through their histories. It is equally true, as
Comte has declared, that 'no conception can
be understood except through its history.'
The latest efforts of human invention, says
Mr. Ferguson, are but a continuation of certain

devices which were practised in the earliest ages of the world. Nor is any human thought so primitive as to have lost its bearing on our own thought, nor so ancient as to have broken its connection with our own life.* The historical method of study is, we are convinced, the only one that can give firm and reliable knowledge. 'The investigator who turns from his modern text-books,' says Professor Tylor, 'to the antiquated dissertations of the great thinkers of the past gains from the history of his own craft a truer view of the relation of theory to fact, learns from the course of growth in each current hypothesis to appreciate its *raison d'être* and full significance, and even finds that a return to older starting-points may enable him to find new paths where the modern track seems stopped by impassable barriers.'†

* Tylor's ' Primitive Culture.'   † *Ibid.*

CHAPTER VI.

MATHEMATICS—SCIENCE.

~ ' Our hazy and inconstant age, fuller of dreams than of poems, of phantasms than imagination, has great need of the clear, accurate eye of mathematics, and of firm hold upon reality.'—JEAN PAUL RICHTER.

' Geometry, if we may be allowed to say so, has a holy divinity of its own, inasmuch as it imposes its various forms and models on creation, and maintains it in existence up to the present day.'—RABANUS.

' Wherever there is a law and system, wherever there is relation and correspondence of parts, the intellect will make its way—will interfuse amongst the dry bones the blood and pulses of life, and create " a soul under the ribs of death." '—DE QUINCEY.

' It is one of the tragi-comic features of human life that the ardent little explorer, looking out with wide-eyed wonder upon his new world, should now and again find as his first guide a nurse, or even a mother, who will resent the majority of his questions as disturbing the luxurious mood of indolence in which she chooses to pass her days. We can never know how much valuable mental activity has been checked, how much hope and courage cast down, by this kind of treatment. Yet, happily, the questioning impulse is not easily eradicated, and a child who has suffered at the outset from this wholesale contempt may be fortunate enough to meet, while the spirit of investigation is still upon him, one who knows, and who has the good nature and the patience to impart what he knows in response to the child's appeal.'
—JAMES SULLY.

IN our educational systems the study of mathematics is pursued almost exclusively for its disciplinary value. Beyond a little elementary arithmetic, how few of us have occasion to utilize the knowledge we have gained from this study in after-life! The rules and propositions that we become familiar with in our school-days soon drop out of our memory, and in after-life we naturally ask ourselves whether all the valuable hours spent in mastering them were not so much time wasted.

It will not be difficult to show that intelligent opinion is on the whole opposed to this view. Dr. Donaldson, a strong advocate of classical training, allows that mathematics, if pursued in moderation and simultaneously with other studies, may correct the habit of mental distraction, and substitute for it the habit of continuous attention. In 'The Advancement of Learning,' Lord Bacon complains that this use of mathematics is ignored, and observes: 'Men do not sufficiently understand the excellent use of the pure mathematics, in that they do remedy and cure many defects in the wit and faculties intellectual. For, if the wit be too dull, they sharpen it; if too wandering, they fix it; if too inherent in the sense, they abstract it.'

In 'A French Eton,' Matthew Arnold says that M. de la Rive, a distinguished Swiss, told him he could trace in the educated class of

Frenchmen a precision of mind distinctly due
to the sound and close mathematical training of their schools.   Dr. Whewell, again,
referring to successful lawyers, remarks the
extraordinary coincidence of professional eminence in after-life with mathematical distinction in their University careers, and from this
concludes that ' our studies may be an admirable discipline and preparation for pursuits
extremely different from our own.'   'That
mathematical habits,' he observes, 'do, or
have done, so much to make men good
lawyers, is not an unimportant consideration
with respect to that profession ; but it is far
more important as showing what such a training may effect in reference to other and wider
studies.'

But what are the peculiar characteristics of
mathematics that tend to such a result ?

' The peculiar character of mathematical
truth,' says Dr. Whewell, ' is that it is necessarily and inevitably true.'   We are concerned
in this study, he proceeds, ' with long chains of
reasoning in which each link hangs from all the
preceding.'   We take for our premises some
obvious truth, and upon this we build a suc-
cession of other truths, which, if our reasoning be sound, are each and all necessarily and
inevitably true.   ' Pure mathematics,' says the
Rev. J. Joyce, ' have one peculiar and distinguishing advantage, that they occasion no

disputes among wrangling disputants as in other branches of knowledge ; and the reason is because the definitions of the terms are premised, and everybody that reads a proposition has the same idea of every part of it. Hence it is easy to put an end to all mathematical controversies by showing either that our adversary has not stuck to his definitions, or has not laid down true premises, or else that he has drawn false conclusions from true principles.'

In mathematics, though concentration is of the first importance—for, says Lord Bacon, in demonstrations, if a man's wits be called away never so little he must begin again—the study is, moreover, fraught with encouragement to the student, for the difficulties mathematics present are such ' as will bend to a resolute effort of the mind ;'* and not only so, for they have ' the additional recommendation that they are apt to stimulate and irritate the mind to make that effort.'†

The disciplinary value of mathematics is admitted, not only by the advocates of a paramount mathematical training, but also by those who give greater prominence to the literary curriculum. We offer an opinion from each side.

In regard to this, the eminent mathematician Dr. Whewell observes : ' The man of mathe-

* De Quincey.   † *Ibid.*

matical genius who, by the demands of his college or university, is led to become familiar with the best Greek and Latin classics, becomes thus a man of liberal education, instead of being merely a powerful calculator. The elegant classical scholar who is compelled in the same way to master the propositions of geometry and mechanics, acquires among them habits of rigour of thought and connection of reasoning. He thus becomes fitted to deal with any subject with which reason can be concerned, and to estimate the prospects of science instead of being kept down to the level of the mere scholar learned in the literature of the past, but illogical and incoherent in his thoughts, and incapable of grappling with the questions which the present and the future offer.'

Dr. Donaldson we may cite as a fair exponent of the other side. In his work on ' Classical Scholarship and Classical Learning' he remarks : ' The full cultivation of the reasoning or logical faculty does no doubt require geometry as well as grammar or logic. . . . I feel assured that, although the classical scholar as such would be ill provided for the full discharge of his important functions, if he were not also to a certain extent at least a mathematician, and though a liberal education would be incomplete if it did not add geometry to its grammatical training, the mere mathematician stands in an infinitely lower position in regard to the cultivation of his intellectual powers than even the merely classical scholar.'

But the value of a mathematical training does not solely depend upon its utility in developing the reasoning powers. It has an even more vital and fundamental importance. It is placed by Comte, says Mr. Lewes, 'in virtue of the principle of his classification at the very head of the scale of sciences. But he regards this vast and important science less as a constituent part of natural philosophy than as the *true and fundamental basis of it ;* and he values it not so much for its own intrinsic truths as for its being the great and most powerful instrument in furthering the progress of science.'*

For, he adds, 'it is by the study of mathematics, and it alone, that we can obtain a just and comprehensive idea of what a science really is. It is in that study we ought to learn precisely the general method always followed by the human mind in its positive researches, for nowhere else are questions resolved so completely and deductions prolonged so far with extreme rigour. It is there, too, that our intelligence has given the greatest proof of its power, since the ideas dealt with are the most entirely abstract possible in positive science. All scientific education which does not commence with this study is therefore and of necessity defective at its foundation.'

* Comte's ' Philosophie Positive,' by G. H. Lewes.

These remarks form a fitting introduction to a brief consideration of the subject of scientific education.

Let us first determine what Science is. ' Information, when it is nothing more,' says Dr. Donaldson, 'merely denotes an accumulation of stray particulars by means of the memory; on the other hand, knowledge is information appropriated and thoroughly matured. . . . And when knowledge extends to a methodical comprehension of general laws and principles, it is called science.'

Science, therefore, is a knowledge of the laws of Nature.

'This knowledge,' says Mr. Lewes, 'is the only rational basis of man's action on Nature. By it he foresees what will be the result of the working of any phenomena left to their own spontaneous activity, and by what modifications he may produce a different result more advantageous to himself. *Science* gives power to *foresee*, and *foreseeing* leads to *action*. . . . Prevision is the characteristic and the test of knowledge. If you can predict certain results, and they occur as you predicted, then you are assured that your knowledge is correct. If the wind blows according to the will of Boreas, we may, indeed, *propitiate* his favour, but we cannot calculate upon it. We can have no certain knowledge whether the wind will blow or not. If, on the other hand, it is subject to laws, like

everything else, once discover these laws, and
men will predict concerning it as they predict
concerning other matters.'

Knowing as we now do the general aim of
scientific education, let us pause to consider
what is the most rational and appropriate
system. If we have hitherto been guided by
Herbert Spencer's dictum that the education
of the child should accord both in mode and
arrangement with the education of mankind,
considered historically, we feel that, in dealing
with scientific education, the justification for
this preference will be the more convincing.

The remarks we made in an earlier part of this
book, when treating of the cultivation of the
faculty of observation, have special application
in determining what the true elementary training in science is. In the first place, the mind
must have material to work upon. During
childhood, the senses are particularly active in
detecting and registering in the mind the
multitude of impressions by which they are
assailed. It is a period peculiarly devoted to
the accumulation of stray knowledge, a process
destined in due course to call into requisition
the power of classification or arrangement
which has not yet appeared. We should at this
time encourage in the child the habit of wholesome curiosity. To repress it is to stifle one of
Nature's most potent intellectual agencies. But
to encourage it is not enough. It needs satis-

faction, and it becomes our duty to minister to it. 'How much better and more intelligent,' says Archdeacon Wilson, 'would early training be if curiosity were looked on as the store of force, the possible love of knowledge in embryo in the boy's mind, which in its later transformations is so highly valued.'

In general science, and particularly natural history, the curiosity of boys is seldom dormant. The subjects have peculiar charm for them, and this charm it is our duty to promote and sustain, that the effect of so ennobling and vitalizing an intellectual activity may not be lost to them in after-life. But not only does the subject-matter of scientific education afford special incitement to the young mind, but, as Mr. Sidgwick says, it teaches him what he will afterwards be more glad to know; it is, in fact, a book which, when once opened, will never be shut up and put by. For, Mr. Sidgwick adds, 'Physical science is now so bound up with all the interests of mankind, from the lowest and most material to the loftiest and most profound; it is so engrossing in its infinite detail, so exciting in its progress and promise, so fascinating in the varied beauty of its revelations, that it draws to itself an ever-increasing amount of intellectual energy; so that the intellectual man who has been trained without it must feel at every turn his inability to comprehend thoroughly the present phase of the progress of humanity, and his limited sympathy with the thoughts and feelings, labours and aspirations, of his fellow-men.'

In one particular, at least, the methods of classical and scientific training are in singular contrast. Whereas the classical student pursues his labours almost entirely through the agency of books, the science student, on the other hand, is concerned with facts, with living facts, which he must observe for himself. 'No teaching of science,' remarks Huxley, 'is worth anything as a mental discipline, which is not based upon direct perception of the facts, and practical exercise of the observing and logical faculties upon them.'

As a mental discipline it would be hard to overrate the value of scientific training; even our very errors, De Quincey says, are full of instruction. 'There is no intellectual discipline,' remarks John Stuart Mill, 'more important than that which the experimental sciences afford. Their whole occupation consists in doing well what all of us, during the whole of life, are engaged in doing for the most part badly.'

Perhaps the importance of this training as a means of developing the mental faculties has never been more emphatically stated than in the Report of the Royal Commission on Education of 1861. The Commissioners say in regard to the study of natural science: 'We believe that its value as a means of opening the mind and disciplining the faculties is recognised by all who have taken the trouble to acquire it, whether men of business or of leisure. It quickens and cultivates directly the faculty of observation, which in very many persons lies almost dormant through life, the power of accurate and rapid generalization, and the mental habit of method and arrangement; it accustoms young persons to trace the sequence of cause and effect; it familiarizes them with a kind of reasoning which interests them, and which they can promptly comprehend; and it is perhaps the best corrective for that indolence which is the vice of half-awakened minds, which shrink from any exertion that is not, like an effort of memory, merely mechanical.'

But the pursuit of science as a form of mental discipline need not comprehend the numerous departments that are necessary when knowledge is sought. Great as is the intrinsic importance of scientific facts, the quest of them should not form the only consideration in school training. Scientific method is what the scholar should be stimulated to attain, and those departments of science should be systematically pursued that most conduce to the realization of this aim. 'There are two kinds of physical science,' says Huxley: 'the one regards form and the relations of forms

to one another; the other deals with causes
and effects. In many of what we term sciences,
these two kinds are mixed up together; but
systematic botany is a pure example of the
former kind, and physics of the latter kind,
of science. . . . Every educational advantage
which training in physical science can give is
obtainable from the proper study of these two.'

We have still to point out the value of
scientific education as a means of imparting
knowledge.

'The most obvious part of the value of
scientific instruction,' says John Stuart Mill,
'the mere information it gives, speaks for itself.
We are born into a world which we have not
made—a world whose phenomena take place
according to fixed laws, of which we do not
bring any knowledge into the world with us.
In such a world we are appointed to live, and
in it all our work is to be done. Our whole
working power depends on knowing the laws
of the world—in other words, the properties of
the things we have to work with, and work
among, and to work upon.'

When science is pursued for the value of the
knowledge it affords, it must be a general
study. We may push our original investiga-
tions further in one department of science than
in others ; but we can do so successfully only
after we have grown to realize the unity and
interdependence of the operations of Nature,

and are able to bring to the explanation of the phenomenon under our consideration the knowledge that can unravel the various contributory causes of which it is the product.

' It can hardly,' says Sir John Herschel, ' be pressed forcibly enough on the attention of the student of Nature, that there is scarcely any natural phenomenon which can be fully and completely explained, in all its circumstances, without a union of several, perhaps of all, the sciences.'

Not only should our science training be general; but it should be methodical. Applying Herbert Spencer's precept, it should advance in conformity with the order in which mankind has itself gained a knowledge of the universe.

' It is at the very root of Comte's system,' remarks Mr. Lewes, ' that until the sciences are learnt in their natural order, which at present is seldom the case, a scientific education will be incapable of realizing its most general and essential results.'

And he gives the following illustration on the basis of Comte's classification of the sciences : ' The natural philosophers who have not in the first place studied astronomy, at least under the general point of view ; the chemists who, before occupying themselves with their own science, have not previously studied astronomy, and, after it, physics ; the physiologists who have not prepared themselves for their special labours by a preliminary study of astronomy, of physics, and of chemistry—all want one of the fundamental conditions of their intellectual development.'

The utility of scientific knowledge is obvious. The whole of our commerce and manufacturing industries are based upon it. As we have wrested from Nature her secrets, so our opportunities of progress in power, in happiness, and in culture, have increased. We have acquired the knowledge of the constitution of the earth's crust only to snatch from its entrails the two prime sources of our great industrial activity—coal and iron : the one the source of light, heat, and mechanical power ; the other the means by which alone our great mechanical knowledge can be practically utilized. We have studied the motions of the heavenly bodies and the mysteries of the magnet, and, with knowledge thus acquired, have mapped out the surface of the globe, and are thus enabled to send our ships to distant parts of the earth with unerring precision. We no longer regard electricity as the ' flashing wrath of the ᐧDeity,' but have learnt to control and utilize it until it has become one of the potent factors in our civilization. Again, from the study of the constitution and functions of man, the art of medicine has arisen, with all its possibilities of arresting and curing our diseases. In fact, there is no department of science which man has not made to answer to his material needs.

But scientific education has a deeper and more significant purpose.* As a spiritual influence, illustrative of the singleness of mind that pervades all creation, its chief importance stands declared. The whole use of astronomy is not included in the pages of the 'Nautical Almanack.' On the contrary, its immediate justification rests in the fact that it has unfolded to us a perfect intelligence manifesting itself in laws and principles of faultless harmony and stability, by which the universe is, we may safely affirm, preserved in perpetual equilibrium. As an individual fact, it may be interesting to know the exact amount of perturbation existing between the planets Uranus and Neptune ; but it is, or should be, of universal interest, and, what is more, the subject of reverential wonder and admiration, the fact that so precise and Providential is the balance of the system that that perturbation will in time receive its due compensation and the stability of the system be preserved.

'To the establishment of invariable laws,' says Spencer, 'we owe our emancipation from the grossest superstitions. But for science we should be still worshipping fetishes, or, with

---

* The succeeding paragraphs appeared substantially as here produced in a contribution of the author's to the *Echo* newspaper.

hecatombs of victims, propitiating diabolical deities.'

By a profound intuition, the ancient Jews maintained the existence of but one God, a belief to which modern science fully subscribes. The entire body of scientific evidence is accumulated in support of it, and every new discovery helps to confirm it. By the aid of science we find that the evolution of humanity is ordered upon a wise and symmetrical design, and that the perplexities that vex us arise only when our knowledge is imperfect, and we view the events of life as isolated facts, and not in perspective in the scheme of the creation.

Thus, it is to attain a knowledge of the laws of Nature that our highest efforts should be consecrated. In doing this, we shall find 'a guide and sanction for our conduct—a sanction no longer external and imposed by the State, but internal and imposed by the mind.'*

* 'Aristotle,' by Thomas Davidson.

CHAPTER VII.

ART.

'If I were to define art, I should be inclined to call it the endeavour after perfection in execution. If we meet with even a piece of mechanical work which bears the marks of being done in this spirit—which is done as if the workman loved it, and tried to make it as good as possible, though something less good would have answered the purpose for which it was ostensibly made— we say that he has worked like an artist.'—MILL.

'The latest efforts of human invention are but a continuation of certain devices which were practised in the earliest ages of the world and in the rudest state of mankind.'—FERGUSON.

'Music is a moral law. It gives a soul to the universe, wings to the mind, flight to the imagination, a charm to sadness, gaiety and life to everything.'—PLATO.

WE have already, in speaking of the cultivation of imagination, approached the subject of art training. Let us now deal with it more in detail.

But, in the first place, what is Art?

In speaking of poetry, Lord Bacon observes: 'The use of this feigned history hath been to give some shadow of satisfaction to the mind of man in those points wherein the nature of things doth deny it, the world being in proportion inferior to the soul; by reason whereof there is, agreeable to the spirit of man, a more ample greatness, a more exact goodness, and a more absolute variety, than can be found in the nature of things.'

And, we may add, there is a more perfect beauty. 'There is nothing of any kind so fair,' says Cicero, 'that there may not be a fairer conceived by the mind. We can conceive of statues more perfect than those of Phidias. Nor did the artist, when he made the statue of Jupiter or Minerva, contemplate any one individual form from which to take a likeness; but there was in his mind a form of beauty, gazing on which he guided his hand and skill in imitation of it.'

To give form to this ideal is the purpose of art. To select and assemble in one whole beauties and perfections which are usually seen in different individuals, excluding everything defective or unseemly, and thereout to form a type or model of the species :* this is the method of art. Zeuxis, it is said, drew his picture of ideal beauty from *five* of the most beautiful women of Crotona.

'Art,' says John Stuart Mill, 'when really cultivated, and not merely practised empirically, maintains, what it first gave the con-

* Dr. Fleming.

ception of, an ideal beauty, to be externally aimed at, though surpassing what can be actually attained.'

From the standpoint of education, art has a twofold aspect, according as it appeals to the intellect or the emotions. It appeals to the intellect in so far as it displays technical skill and craftsmanship. It appeals to the emotions by reason of the sentiments it illustrates. 'Now, an ignorant man,' says M. Chesneau, 'is not insensible to the influence of a work of art, when it is conceived by so high a standard as to rouse the best impulses of our nature. Such works as these he at once apprehends.' 'His heart,' says George Sand, 'will teach him what his ignorance hides from him.' . . . 'It is the feeling of the work that appeals to him, and he will be no less, or perhaps even more, touched by a piece of indifferent execution as by learned workmanship.'

But, adds M. Chesneau, he is insensible to the 'cleverness, solidity, or technical skill which enchant a practised amateur,' and the 'subtlety, refinement, and special skill of handling will be a sealed book to him.'

This distinction seems to indicate the proper limitations of art training in our schools. We have already, when treating of the imagination, called attention to the injurious effect that is likely to ensue from an untimely stimulation of the emotions. What Jean Paul Richter so

eloquently prescribes for the healthy cultivation
of the poetic genius may be taken to apply to
the art temperament in general. 'The feelings
of the poet,' he says, 'should be closely and
coolly covered, and the hardest and driest
sciences should retard the bursting blossoms
till the due springtime.' But the feelings of
the poet differ from the feelings of the painter
or musician only in their instrument of expression: the one has language, the others colour
and tone. All are a means to an end, and the
end in each is the expression of feeling.

But though systematic expression of feeling
in youth is to be discountenanced, it does not
justify the total exclusion of art training from
our curriculum. It must be allowed that the
technical side of art—the acquisition of art
knowledge and skill in art method—is not open
to the objection we have named. On the contrary, it is disciplinary, and, moreover, necessary as an aid to subsequent general culture.
But let us give this more careful consideration.

'The beginnings of every study,' says Quintilian, 'are formed in accordance with some
prescribed rule.' The accumulated experience
of mankind, sifted and refined by frequent re-
examination, becomes embodied in rules and
principles. These, by general assent, have
become the foundations of knowledge. They
are steps to higher efforts, and must be scaled

before the mind can achieve original work. 'Though to invent,' says the same author, 'was first in order of time, and holds the first place in merit, yet it is of advantage to copy what has been invented with success. Indeed, the whole conduct of life is based on the desire of doing ourselves that which we approve in others.' Thus it is that the beginning of art consists of imitation.

The art training of our children, then, should be confined to the imitation of what has already been invented with success, to the pursuit of established principles and method. After they have acquired these, after they have 'equalled what they imitate,' they can, when relaxation in other studies is permissible, push on to higher efforts. 'The pupil must first obtain a thorough knowledge of art,' says Mr. Val Prinsep, 'and then render his own feelings, which he alone can find out.' By this the mind will receive its proper direction of impulse, and will be the less liable to waste its efforts in vague imaginings.

The question then arises, To what limit should this art training be followed? By the majority of us, art will not be studied at school with a view to pursuing it in after-life. We shall relinquish its pursuit to those who are specially endowed. But however we decide, we shall have reaped advantages from our training that will endure through life; we shall

have it in our power to become judges of what
is excellent in art, and take a proper pleasure
therein ;* and this power acquired, we shall
imperceptibly grow, not only to seek perfection
in art, but to idealize, as much as possible,
every work we do, and, most of all, our own
characters and lives.†

In dealing with music as a subject of school
education, we are brought face to face with
several important considerations.

What we have already said of art generally
applies particularly to music. Music is, perhaps
more than poetry or painting, the language of
the emotions, and is, therefore, particularly
open to the objection that we made when
treating of the emotional side of art. Moreover, as music is more commonly studied and
is pushed to a more advanced stage in school
years than other branches of art, the immode-
rate pursuit of music lays itself the more open
to condemnation.

Musicians, when suffered to become absorbed
from youth upward in their profession, not un-
frequently manifest the most restricted general
knowledge, and the most limited general capacity. Mr. Haweis, while admitting the perils
that attend the musician's mental development, attributes their origin to the following
cause :

' They ' (the musicians), he observes, ' have

* Aristotle. † Mill.

8

not so much time for reading and thinking. . . .
The practice of musical mechanism is not intellectual: it does not nourish the brain or
feed the heart; it does not even leave the
mind at liberty to think; it chokes everything
but its own development, and that is merely
a physical development. . . . The musician's
strict exercise—which, after all, takes up a
great deal of his time—admits of very little
intellect, imagination, or emotion. It requires
industry, perception, and nerve; in short, because it is more mechanical, it is therefore less
refining and elevating.'

In our next chapter it will be shown how
morality lies in the co-ordinate development
of all the faculties, and how the forcing of one
or more at the expense of the rest not unfre-
quently results in an injurious disturbance of
the balance of the moral nature. Upon this
thesis it will be seen how harmful must be
the effect of the dedication of a child's best
efforts to the paramount pursuit of music.
The science learnt in early years too quickly
resolves itself into the art. The art gradually
appropriates more and more of the pupil's
intellectual activity, until in the end it exercises
a supreme dominion over his mental faculties,
and absorbs, for the purposes of its own development, that energy which should be exercised towards the maturing of the mind as a
whole.

Further, though, as we are reminded by Mr. Haweis, the musician's strict exercise is merely physical development, it is not exercise of a kind capable of counteracting the deleterious effects of the pursuit of music unaccompanied by that more beneficial exercise so necessary for the preservation of the health of both body and mind. The effect, says Plato, on one who studies music exclusively, is that, 'should he possess any spirit, it softens it like iron, and makes it serviceable instead of useless and harsh. When, however, he positively declines desisting, and becomes the victim of a kind of fascination, after this he is melted and dissolved, till his spirit is quite spent, and the nerves are, as it were, cut out from his soul.'

But let us not be understood from the above to depreciate the value of music, nor to discountenance the pursuit of it. We do neither. With due regard to the claims of other studies, we consider the study of music necessary to insure the full development of the mental faculties; for we hold that a mind insensible to the beauties of art, and therefore of music, is one that is but half awakened. What is concerned with music, says Plato, ought, somehow, to terminate with the love of the beautiful. It is not enough that the mind be transformed into the 'clear, cold logic-engine' of the scientist or man of business. The emotions are as much a fact as the judgment, and each expresses itself in its own way, and has need of its own particular language. By all means, let us study music ; but beware, lest it become our infatuation !

CHAPTER VIII.

THE BASIS OF MORALITY.

'If, then, intellect is something divine in relation to man, the life lived according to it must be divine in relation to human life.'—ARISTOTLE.

'Right action is better than right knowledge ; but, in order to do what is right, we must know what is right.'—BALUZIUS.

> 'Refrain to-night ;
> And that shall lend a kind of easiness
> To the next abstinence : the next more easy :
> For use almost can change the stamp of nature,
> And either curb the devil, or throw him out
> With wondrous potency.'
>
> SHAKESPEARE.

'But men must know that in this theatre of man's life it is reserved only for God and angels to be lookers-on.'—BACON.

IN the 'Republic' Plato has said that it is incumbent upon each one of us to learn and find out who will make him expert and intelligent to discern a good life and a bad.

To choose the good and reject the bad ; to practise virtue, that 'health, beauty, and good habit of the soul,' and eschew vice, 'its disease, deformity, and infirmity'—this is the final aim

of education: knowledge, to ˙ point the way;
and discipline, to reach the goal.

And how is this perfection to be gained?
Is virtue a separate faculty, like the memory,
that can be developed by artificial aids?
'Perfection,' Matthew Arnold says, 'is a harmonious expansion of *all* the powers which
make the beauty and worth of human nature,
and it is not consistent with the over-develop-
ment of any one power at the expense of the
rest.' It is 'the fit details strictly combined in
view of a large general result.'

But though perfection lies in the harmonious
expansion of all the powers which make the
beauty and worth of human nature, we must
not from that conclude that perfection lies in
their equal development. It is, no doubt,
possible for us to conceive a type of human
nature manifesting to the full our ideals of perfection both in its collective and separate
features; but so far is such a type beyond our
hopes of realization that we must be content to
regard perfection as a relative term, and, for
the purposes of education, hold each individual
perfect whose unequal powers for good have
expanded to *their* utmost without disturbing the
general harmony of the whole. We must not
seek perfection in uniformity. No two flowers,
even of the same species or on the same plant,
are alike. Though corresponding in their
general characteristics, there are diversities of

form, colour, and scent that give to each its individuality. We must attempt to unfold, says Lavater, only what Nature is desirous of unfolding, give what Nature is capable of receiving, and take away that with which Nature would not be encumbered.

The form of education most calculated to enlarge the scholar's special power is, unfortunately, not that best suited to promote his growth of character. In other words, the concentration that is necessary for success in life will operate to disturb the harmony and balance of the mind upon which all true morality is based.

'Is it better,' says De Quincey, 'to be a profound student or a comprehensive one? In some degree,' he answers, 'this must depend upon the direction of the studies; but generally, I think, it is better for the interests of knowledge that the scholar should aim at profundity, and better for the interests of the individual that he should aim at comprehensiveness. A due balance and equilibrium of the mind is best preserved by a large and multiform knowledge; but knowledge itself is best served by an exclusive (or at least paramount) dedication of one mind to one science.'

An equivalent, but perhaps more suggestive, answer to this question is made by Schiller in 'The Æsthetic Education of Man.' He there says: 'Partial exercise of the faculties leads the individual undoubtedly into error, but the species into truth.' And further on : 'Extraordinary men are formed by energetic and over-excited spasms, as it were, in the individual faculties, though it is true that the equable exercise of all the faculties in harmony with each other can alone make happy and perfect men.'

We must not ignore the fact that the conditions of present life tend more and more to make specialists of us. 'So vast is the accumulation of facts,' observes Spencer, 'which men of science have before them, that only by dividing and subdividing their labours can they deal with it.' It is as when we visit some great cathedral, which, as we approach, we regard as a whole, contemplating the harmony of its proportions and the elegance of its design. It is not until we are under its very towers that the charm of its detail strikes us, and we then give ourselves over to an examination of its subordinate beauties. The closer we approach, the better are we able to distinguish its minor characteristics; but the less able are we to realize the proportion and relationship of its various parts. Were it not for the aid of memory, we should lose all appreciation of the symmetry of the building as a whole, and be only too ready to attach to the particular part under our immediate observation a greater or less importance than is justified by the facts.

' Each,' said Plato, 'may exercise one business well, but many not.' This is no doubt true ; and it is equally true that it is the duty of each one of us to exercise his business well. But we may here apply the advice given by Sir James Sawyer in an address to medical students, when he insisted that they should aim at becoming specialists rather in practice than in knowledge. Our knowledge should not be confined to merely what is necessary for excellence in our particular vocation ; for in that case we shall falsify, through the habitual frame of mind this exclusive cultivation will give us, every other impression our minds will receive.*

Every pursuit, when followed exclusively, carries its own Nemesis with it. ' Experience proves,' says John Stuart Mill, ' that there is no one study or pursuit which, practised to the exclusion of all others, does not narrow and pervert the mind, breeding in it a class of prejudices special to that pursuit, besides a general prejudice, common to all narrow specialities, against large views, from an incapacity to take in and appreciate the grounds of them.'

Of the attorney of his day, Bishop Earle wrote : ' His business gives him not leave to think of his conscience, and when the time or term of his life is going out, for doomsday he is secure, for he hopes he has a trick to reverse judgment.'

A caricature, no doubt, but only a partial

* Herbart.

one! We have here a mind 'narrowed and perverted' by the prejudices and illusions peculiar to a particular calling. The broader principles of justice and honesty are obscured, and conduct is shaped by the exigencies of professional interests.

Medicine, too, is open to the same censure. The practice of vivisection can, we feel sure, prevail only where the mind, devoted to one pursuit, has ceased to recognise the higher claims of pity. Little is of value to such a mind but what is subservient to the advancement of its one aim, and life itself becomes merely a temporary vehicle for exhibiting in prolonged suffering its obscure phenomena.

We have already shown the inadvisability of introducing special studies into the general education. The reason for this will now be more apparent. 'The individuality,' says Herbart, 'must first be changed through widened interest, and approximate to a general form, before they (the teachers) can venture to think they will find it amenable to the general obligatory moral law.'

Morality, he contends, has its root in manysidedness. 'The more individuality is blended with manysidedness,' he observes, 'the more easily will the character assert its sway over the individual.' For morality lies in our power to estimate the due proportion and value of circumstances and things, and to make the exercise of this power the moving impulse of our life ; and it is at variance with our 'inaptitude for seeing more than one side of a thing, and our intense energetic absorption in the particular pursuit we happen to be following.'*

But general knowledge is not necessarily surface knowledge—a 'versatility in everything, with sure knowledge of nothing.' Sokrates has said that 'it is better to accomplish a little thoroughly than a great deal insufficiently.' But it is one thing, Sir John Lubbock has reminded us, to know a few stray facts of a subject ; it is quite a different thing to be well grounded in it.

The limit of general studies has been well defined by John Stuart Mill, and, helping as it does to disperse the mists that envelop a difficult subject, we cannot do better than quote his words. He observes : ' To have a general knowledge of a subject is to know only its leading truths, but to know these not superficially, but thoroughly, so as to have a true conception of the subject in its great features, leaving the minor details to those who require them for the purposes of their special pursuit. There is no incompatibility between knowing a wide range of subjects up to this point, and some one subject with the completeness required by those who make it their principal occupation. It is this combination which gives an

* Matthew Arnold.

enlightened public—a body of cultivated intellects, each taught by its attainments in its own
province what real knowledge is, and knowing
enough of other subjects to be able to discern
who are those that know them better.'

But such knowledge is not of itself the final
guide to conduct. We need the help of the
imagination. We must cultivate standards of
excellence, ideals of perfection. We can have
no direct experience of absolute goodness or
perfect beauty; but, from familiarity with objects
which display the qualities of goodness and
beauty, we can form our own idea of absolute
goodness and perfect beauty, and these will be
our standards and exemplars in the work of
self-examination.

'The ideal,' says Dr. Fleming, 'is to be
attained by selecting and assembling in the
one whole the beauties and perfections which
are usually seen in different individuals, excluding everything defective or unseemly, so as
to form a type or model of the species. Thus,
the Apollo Belvedere is an *ideal* of the beauty
and proportion of the human frame; the
Farnese Hercules is an *ideal* of manly strength.
The ideal can only be attained by following
Nature. There must be no elements nor combinations but such as Nature exhibits; but the
elements of beauty and perfection must be dis-
engaged from individuals, and embodied in one
faultless whole.'

Hence it will be seen that our ideals correspond with the state of our mental cultivation. The Apollo Belvedere does not exist in Nature. It is the embodiment of the sculptor's acquaintance with many separate instances of manly beauty. The wider his experience has been, the more likely is his work to show the influence of all available instances of manly beauty. His ideal cannot be affected by instances of manly beauty of which he is not cognizant.

So it is with our ideals of justice, duty, and the like. They will be high according to the extent of our acquaintance with individual instances of justice and duty, and we shall shape our composite idea of perfect justice and perfect duty from the instances of individual justice and individual duty that come under our notice. As we shall regard our own imperfections by the light of our idea of perfection, how necessary it is that our idea should be a high one : for it is our duty, says Aristotle, as far as may be, to act as immortal beings, and do all we can to live in accordance with the supreme part of us.

But let us not indulge the vain belief that so many have held, that the intellectual conditions of morality are our only concern. To cite the parallel that Sokrates so frequently employed of the education of the professional man—the surgeon, for instance. As we have shown in an earlier chapter, a full knowledge of the human body is not identical with skill in surgery; for, though skill in surgery cannot be attained until this previous knowledge be acquired, we know by experience that the intellectual conditions of surgery do not meet the whole demand. In like manner, the knowledge of what is right in conduct is not identical with the successful practice of what is right. No two ideas are more frequently confounded in the unscientific mind than science and art, theory and practice. To keep them distinct is to render intelligible much that is apparently obscure in the material of our thought. Amongst other subjects upon which the observance of this precaution sheds especial light is that of which we now treat, moral training.

There is a science of good conduct, and there is an art of good conduct—the theory and practice respectively. Our remarks hitherto, devoted as they have been to maintaining the importance of knowledge in building up the moral nature, have touched upon only the theory of conduct. But as the surgeon would be but half equipped if he possessed only the knowledge of anatomy and physiology without skill in practice, so, in matters of conduct, the individual is ill furnished if he have only acquired the knowledge of what is right in conduct without the habit of putting it in practice.

Habits, as we have shown, are acquired by virtue of things done, not things known. As practice is necessary to give skill in surgery, so the practice of good acts can alone create good habits. But the surgeon's practice is regulated by method, and his method is determined by his knowledge of that upon which he has to operate. In like manner, the development of our moral habits should proceed with method, and the method should be determined by the general knowledge that we have shown to constitute the theory of moral training.

But the greater the experience of the surgeon, the more intimate is the alliance between theory and practice, and the less easily can he be induced to extend his practice beyond the limits of his knowledge, or, conversely, the less likely is he to allow his knowledge to fall short of the limits of his practice. In matters of conduct, may we not safely affirm that the more the knowledge of what is right underlies the habit of doing what is right, the more potent will knowledge relatively become in controlling the moral nature? We are, of course, aware that the professional man has inducements to preserve his fitness undiminished, that find only a partial parallel in the domain of ethics. The stimulus of competition, the sanctity of professional reputation, the material remuneration of successful practice, are considerations that have weight, even in natures in which the pride of good workmanship brings its own reward. For all this, however, it will be found that the individual, even apart from outside inducements, has a singular propensity for adjusting his acts to the dictates of his reason, and the more so the less casual and indefinite the knowledge is upon which his reason operates.

But the acquisition of knowledge is a slow process, and habits are formed only after long practice. What, then, is to be the moral force of childhood and youth ? It is a question that gives rise to much speculation, and to which no satisfactory answer is obtainable. Each has his pet nostrum, the product of 'much faith and much chance'; but none seem able to refer their method to an adequate knowledge of the material with which they have to deal. The only definite answer given to this question is that supplied by the youth of the race. From this comparison we learn, if we take the history of the ancient Greeks as our guide, that in the youth of the race feeling is the paramount moral sanction. We also learn that feeling gradually gives place to judgment, and that during this period of transition the moral force is external and imposed, destined, it would seem, to assist the formation of moral habits at a time when the knowledge acquired is inadequate as a guide.

It remains only to observe that work is one of the chief factors in moral training. 'Religion without work,' says Froebel, 'runs the risk of becoming empty dreaming, passing enthusiasm, and an evanescent phantom, as work without religion makes man a beast of burden or a machine.' The acquisition of knowledge and the development of power must be regarded as only a means to an end; and that end is action. 'But for contemplation,' observes Lord Bacon, 'which should be finished in itself, without casting beams upon society, surely divinity knoweth it not.'

CHAPTER IX.

TEACHERS.

'People think that morals are corrupted in schools: for indeed they are at times corrupted; but such may be the case even at home. Many proofs of this fact may be adduced, proofs of character having been vitiated, as well as preserved with the utmost purity, under both modes of education. It is the disposition of the individual pupil, and the care taken of him, that makes the whole difference.'—QUINTILIAN.

'Those who, in eagerness to cultivate their pupils' minds, are reckless of their bodies, do not remember that success in the world depends more on energy than information, and that a policy which in cramming with information undermines energy is self-defeating.'—SPENCER.

'The beginning is more than half the whole.'—*From the Greek.*

'It is a singular circumstance that within a quarter of a mile of the well-head of the Wye arises the Severn. The two springs are nearly alike; but the fortunes of rivers, like those of men, are owing to various little circumstances of which they take advantage in the early part of their course.'—WILLIAM GILPIN.

MUCH has been written and said upon the respective advantages of public and private tuition. Parents who hope, by the engagement of a private tutor, to shield their children from the possibility of contamination in school life, often have reason to regret the loss of the virile influence that the school competition and sports afford. Dupanloup once said: ' I have heard a man of great sense utter this remarkable word, " If a usurping and able Government wanted to get rid of great races in the country, and root them out, it need only come down to this, that it require of them, out of respect for themselves, to bring up their children at home, alone, far from their equals, shut up in the narrow horizon of a private education and a private tutor." '*

The importance of bodily exercise as a concomitant to mental effort is well recognised in the public schools of this country. In ancient Greece it was an important feature in the training of the young, regarded as it was not only as a means towards bodily development, but also as an indispensable aid in promoting strength and vigour of mind.

Gymnastics was the form of exercise employed by the children of ancient Greece. They were trained to acquire ' the grace and activity of motion, the free step, the erect mien, the healthy glow.'† Mr. Mahaffy, in his ' Old Greek Education,' contrasts this practice with the advantages enjoyed by the modern English

* Rev. Thomas Hughes' ' Loyola.'
† Dr. Donaldson.

scholar, and observes : 'I say it quite deliberately : the public schoolboy, who is trained in
cricket, football, and rowing, and who in his
holidays can obtain riding, salmon-fishing,
hunting and shooting, enjoys a physical training
which no classical days ever equalled.' Froebel,
too, has remarked the profound mental discipline that arises from well-organized school
sports. 'I studied the boys' play,' he says,
'the whole series of games in the open air,
and learned to recognise their mighty power to
awaken and to strengthen the intelligence and
the soul as well as the body.'

All this will be lost to the student who pursues his studies privately. The choice is a
difficult one to make, and has disturbed the
minds of men the most capable of coming to a
decision. Dr. Arnold, in a letter to Sir Thomas
Paisley, gives expression to conflicting views
on the subject. 'The difficulties of education,'
he says, 'stare me in the face whenever I look
at my own four boys. I think by-and-by I
shall put them into the school here, but I shall
do it with trembling. Experience seems to
point out no one plan of education as decidedly
the best; it only says, I think, that public
education is the best where it answers. But,
then, the question is, Will it answer with one's
own boy ? and if it fails, is the failure complete?
It becomes a question of particulars. A very
good private tutor would tempt me to try

private education, or a very good public school, with connections amongst the boys at it, might induce me to venture upon public. Still, there is much chance in the matter; for a school may change its character greatly, even with the same master, by the prevalence of a good or bad set of boys, and this no caution can guard against. But I should certainly advise anything rather than a private school of above thirty boys. Large private schools, I think, are the worst possible system; the choice lies between public schools and an education whose character may be strictly private and domestic.'

With regard to the teachers themselves, when it lies in our power, as it seldom does, to exercise some choice, the following considerations deserve attention.

'The beginning of every work,' says Plato, 'is the most important, especially to anyone young and tender, because then that particular impression is most easily instilled and formed which anyone may wish to imprint on each individual.'

How frequently we hear it remarked that the rudiments of education, because they appear simple, can be taught by the half-educated! 'Let me assure you,' says Huxley, 'that that is the profoundest mistake in the world. There is nothing so difficult to do as to write a good elementary book, and there is nobody so hard to teach properly and well as people who know nothing about a subject. . . . It involves that difficult process of knowing what you know so well that you can talk about it as you can talk about your ordinary business. A man can always talk about his own business. He can always make it plain; but if his knowledge is hearsay, he is afraid to go beyond what he has recollected, and put it before those that are ignorant in such a shape that they shall comprehend it.'

Quintilian has much to say on this topic, and what he has said, like almost everything he has said, deserves attention. Though he is treating primarily of the education of the orator, yet the remarks we now quote apply to education generally.

'Would Philip, King of Macedonia,' he exclaims, 'have wished the first principles of learning to be communicated to his son Alexander by Aristotle, the greatest philosopher of that age, or would Aristotle have undertaken that office, if they had not both thought that the first rudiments of instruction are best treated by the most accomplished teacher, and have an influence on the whole course ?'

'The ablest teachers,' he explains, 'can teach little things best if they will: first, because it is likely that he who excels others in eloquence has gained the most accurate knowledge of the means by which men attain eloquence;

secondly, because method, which with the best-qualified instructors is always plainest, is of great efficacy in teaching; and, lastly, because no man rises to such a height in greater things that lesser fade entirely from his view. . . . What shall be said, too, if it generally happens that instructions given by the most learned are far more easy to be understood and more perspicuous than those of others? . . . The less able a teacher is, the more obscure will he be.'

If we seek enlightenment on this topic from the history of nations, we shall find a most instructive lesson presented to us in the influence exerted on the national mind and character of the ancient Greeks by its great men; for instance, by Lykurgus in Sparta, and by Solon and Kleisthenes in Athens. There can be no doubt that the peculiar qualities of each of these communities were attributable in great measure to the influence of these early law-givers. 'That the Spartans,' says Mr. Grote, 'had an original organization and tendencies common to them with the other Dorians we may readily conceive; but the Lykurgean constitution impressed upon them a peculiar tendency which took them out of the general march.'

But even more profitable as a pattern of sound and judicious culture is the instance afforded by Periklês, who, according to Mr.

Grote, found the Athenian character with 'very marked positive characteristics and suscepti- bilities,' and who, too wise to endeavour to mould it afresh, exercised his talents in bring- ing out and improving its most estimable qualities.

In the foregoing pages we have said much with regard to the aim of education and the mode of attaining it. But systems are of little value if they be not applied by intelligent minds. ' Methods of teaching are very important, but the teacher is of far more importance; and no teaching of these or any other subject is likely to be worth much unless the teacher is thoroughly master of his work, has made it his own by viewing it in various lights, and is independent of any text-book or any order of viewing Nature. He cannot be too discursive in his reading or varied in his attainments; and if he is further able to be prosecuting some original work, however humble, in which his pupils can assist him, they will learn more of the true scientific spirit by contagion than they will gather from the most eloquent lectures.'*

Dr. Arnold holds that a man is only fit to teach so long as he himself is learning daily. ' If the mind once becomes stagnant,' he says, ' it can give no fresh draught to another mind; it is drinking out of a pond, instead of from a spring. And whatever you read tends generally

* Archdeacon Wilson.

to your own increase of power, and will be felt by you in a hundred ways hereafter.'

But great as is the importance of wide knowledge and skill in teaching in a master, they are not so imperative as rectitude of conduct and loftiness of idea. Pupils will insensibly imitate their teachers, and imitations, says Plato, if from earliest youth onwards they be long continued, become established in the manners and natural temper, both as to body and voice and intellect, too.

'There is nothing,' says John Stuart Mill, ' which spreads more contagiously from teacher to pupil than elevation of sentiment. Often and often have students caught from the living influence of a professor a contempt for mean and selfish objects, and a noble ambition to leave the world better than they found it, which they have carried with them throughout life.'

Let us conclude our observations with regard to the qualities necessary in an enlightened teacher by some valuable remarks of the luminous Quintilian : ' Let him adopt, then, above all things, the feelings of a parent towards his pupils, and consider that he succeeds to the place of those by whom the children were entrusted to him. Let him neither have vices in himself nor tolerate them in others. Let his austerity not be stern, nor his affability too easy, lest dislike arise from the one or contempt from the other. Let him discourse frequently on what is honourable and good ; for the oftener he admonishes, the more seldom will he have to chastise. Let him not be of an angry temper, and yet not a conniver at what ought to be corrected. Let him be plain in his mode of teaching and patient of labour, but rather diligent in exacting tasks than fond of giving them of excessive length. Let him reply readily to those who put questions to him, and question of his own accord those who do not. In commending the exercises of his pupils, let him be neither niggardly nor lavish ; for the one quality begets dislike of labour, and the other self-complacency. In amending what requires correction, let him not be harsh, and, least of all, not reproachful ; for that very circumstance, that some tutors blame as if they hated, deters many young men from their proposed course of study. Let him every day say something, and even much, which, when the pupils hear, they may carry away with them ; for though he may point out to them, in their course of reading, plenty of examples for their imitation, yet *the living voice*, as it is called, feeds the mind more nutritiously, and especially the voice of the teacher, whom his pupils, if they are but rightly instructed, both love and reverence. How much more readily we imitate those whom we like can scarcely be expressed.'

Truly a noble ideal! Would that it could

more frequently be realized ! But we fear that
many will say, as Milton did of the qualities he
judged necessary to the teacher, 'that this is
not a bow for every man to shoot with that counts
himself a teacher, but will require sinews almost
equal to those which Homer gave Ulysses.'

THE END.

Elliot Stock, Paternoster Row, London.

www.ingramcontent.com/pod-product-compliance
Lightning Source LLC
Chambersburg PA
CBHW021813190326
41518CB00007B/568